Existence and Spatial Decay of Periodic Navier–Stokes Flows in Exterior Domains

Vom Fachbereich Mathematik
der Technischen Universität Darmstadt
zur Erlangung des Grades eines
Doktors der Naturwissenschaften
(Dr. rer. nat.)
genehmigte

Dissertation

von

Thomas Walter Eiter, M. Sc.

aus

Gelnhausen

Referent: Prof. Dr. Reinhard Farwig
1. Korreferent: Prof. Dr. Mads Kyed
2. Korreferent: Prof. Dr. Giovanni P. Galdi

Darmstadt 2020

Bibliographic information published by the Deutsche Nationalbibliothek

The Deutsche Nationalbibliothek lists this publication in the
Deutsche Nationalbibliografie; detailed bibliographic data are available
in the Internet at http://dnb.d-nb.de .

zugl.: Darmstadt, Technische Universität Darmstadt, Dissertation - D17
Tag der Einreichung: 12.12.2019
Tag der mündlichen Prüfung: 27.02.2020
URN urn:nbn:de:tuda-tuprints-116294
URI https://tuprints.ulb.tu-darmstadt.de/id/eprint/11629

ISBN 978-3-8325-5108-7

Logos Verlag Berlin GmbH
Comeniushof, Gubener Str. 47,
10243 Berlin
Tel.: +49 030 42 85 10 90
Fax: +49 030 42 85 10 92
INTERNET: http://www.logos-verlag.de

Acknowledgment

I would like to express my gratitude to all people who supported me in the completion of this thesis and during my time as a PhD student.

I express my warmest gratitude to Prof. Dr. Mads Kyed, who has been a great mentor since my master's thesis. Although he left Darmstadt during my time as a PhD student, he kept to be a valuable adviser in any concern. We had many fruitful discussions, in person and by phone, and I am thankful for his guidance during the last years and his warm hospitality during my visits in Flensburg.

I owe special thanks to Prof. Dr. Giovanni P. Galdi for the nice hospitality during my stays at the University of Pittsburgh and for refereeing my thesis. We had rich discussions that contributed to the initiation and completion of many projects.

Deepest gratitude is owed to Prof. Dr. Yoshihiro Shibata for the time we spent with many intensive discussions. He invited me several times to Waseda University and was a great host.

Special thanks is due to Prof. Dr. Reinhard Farwig, who gave valuable advice during my time in Darmstadt. His seminars provided me with a fundamental background in the field of fluid dynamics. I thank him for his service as a referee of my thesis.

I also thank my (former) office mates Aday Celik and Johannes Ehlert as well as the whole working group Analysis for the mathematical and everyday conversations that made these last years the enjoyable and productive experience they were. In particular, I express my gratitude to my colleagues Dr. Björn Augner, Sebastian Bechtel, Aday Celik, Klaus Kreß, Jens-Henning Möller, Andreas Schmidt, and Marc Wrona for the valuable proofreading and their useful advice on how to improve this thesis.

Last but not least, I express my gratitude to my friends and family. They supported me in every non-mathematical respect and provided the necessary distraction when I needed it.

Abstract

Consider a rigid body that performs a prescribed motion through an infinite container without boundaries that is filled with a viscous incompressible fluid. The fluid flow around the body is governed by the Navier–Stokes equations

$$\begin{cases} \partial_t u + \eta \wedge u - (\xi + \eta \wedge x) \cdot \nabla u + u \cdot \nabla u = f + \Delta u - \nabla \mathfrak{p} & \text{in } \mathbb{R} \times \Omega, \\ \operatorname{div} u = 0 & \text{in } \mathbb{R} \times \Omega, \\ u = \xi + \eta \wedge x & \text{on } \mathbb{R} \times \partial\Omega, \\ \lim_{|x| \to \infty} u(t, x) = 0 & \text{for } t \in \mathbb{R}. \end{cases} \quad \text{(NSE)}$$

Here Ω is the exterior of the body, u and $\mathfrak{p}$ are velocity and pressure fields of the fluid, and f is a given external force. Moreover, ξ and η describe (time-dependent) translational and rotational velocities of the body. Object of investigation of the present thesis is the configuration where the fluid flow is time-periodic. The first part of this thesis is dedicated to the existence of time-periodic solutions to (NSE), and the second part addresses spatially asymptotic properties of such solutions.

In Chapter 3 we consider the case where the body performs a translation with constant velocity and no rotation. The analysis is based on a suitable linearization of (NSE) given by

$$\begin{cases} \partial_t u - \Delta u - \lambda \partial_1 u + \nabla \mathfrak{p} = f & \text{in } \mathbb{R} \times \Omega, \\ \operatorname{div} u = 0 & \text{in } \mathbb{R} \times \Omega, \\ u = 0 & \text{on } \mathbb{R} \times \partial\Omega \end{cases} \quad \text{(LNSE)}$$

for $\lambda > 0$. We establish new well-posedness results for both steady-state and time-periodic strong solutions to (LNSE) in an exterior domain $\Omega \subset \mathbb{R}^n$, $n \geq 2$, which are then employed to show existence of steady-state and time-periodic solutions to the corresponding nonlinear system under the assumption of "small" data. While solutions to these exterior-domain problems are usually established in a framework of homogeneous Sobolev spaces, the novelty of the presented approach is that it yields a framework where solutions are established in a full Sobolev space.

In Chapter 4 we study problem (NSE) in a three-dimensional exterior domain Ω under suitable assumptions on the data. The associated linearization is examined in a framework of absolutely convergent Fourier series. Since the corresponding resolvent problem is ill-posed in classical Sobolev spaces, we establish a linear theory in homogeneous Sobolev spaces. These new results are applied to obtain existence of time-periodic solutions to the nonlinear system (NSE) for "small" data.

In Chapter 5 we investigate the linear problem (LNSE) in the whole space $\Omega = \mathbb{R}^n$ for general dimension $n \geq 2$ and for general $\lambda \in \mathbb{R}$, which we formulate as a problem on the locally compact abelian group $\mathbb{T} \times \mathbb{R}^n$. We introduce time-periodic fundamental solutions for the solution $(u, \mathfrak{p})$ to these equations. In addition, we develop the concept of a time-periodic fundamental solution for the vorticity field $\operatorname{curl} u$. We investigate integrability properties and show pointwise estimates of these fundamental solutions.

The subject of Chapter 6 is the investigation of asymptotic properties of time-periodic solutions to (NSE) in the case of non-vanishing mean translation velocity. The velocity field is decomposed into a time-independent part and a time-periodic remainder, and we derive pointwise estimates for these parts separately. In this way, new asymptotic properties of both the velocity and the vorticity are discovered.

Zusammenfassung in deutscher Sprache

Man betrachte einen starren Körper, der eine vorgeschriebene Bewegung durch einen unendlichen Behälter ohne Ränder, welcher mit einer viskosen inkompressiblen Flüssigkeit gefüllt ist, vollzieht. Die Flüssigkeitsströmung um den Körper ist bestimmt durch die Navier-Stokes-Gleichungen

$$\begin{cases} \partial_t u + \eta \wedge u - (\xi + \eta \wedge x) \cdot \nabla u + u \cdot \nabla u = f + \Delta u - \nabla \mathfrak{p} & \text{in } \mathbb{R} \times \Omega, \\ \operatorname{div} u = 0 & \text{in } \mathbb{R} \times \Omega, \\ u = \xi + \eta \wedge x & \text{auf } \mathbb{R} \times \partial\Omega, \\ \lim\limits_{|x|\to\infty} u(t,x) = 0 & \text{für } t \in \mathbb{R}. \end{cases}$$

(NSG)

Hierbei sei Ω der Außenraum um den Körper, u und $\mathfrak{p}$ seinen Geschwindigkeitsfeld und Druckfunktion der Flüssigkeit und f sei eine gegebene äußere Kraft. Des Weiteren beschreiben ξ und η die (zeitabhängige) Translations- bzw. Rotationsgeschwindigkeit des Körpers. Der erste Teil dieser Arbeit ist der Existenz von zeitperiodischen Lösungen von (NSG) gewidmet und der zweite Teil befasst sich mit räumlich asymptotischen Eigenschaften solcher Lösungen.

In Kapitel 3 betrachten wir den Fall, wenn der Körper eine Translation mit konstanter Geschwindigkeit und keine Drehung vollzieht. Die Untersuchung basiert auf einer geeigneten Linearisierung von (NSG), die gegeben ist durch

$$\begin{cases} \partial_t u - \Delta u - \lambda \partial_1 u + \nabla \mathfrak{p} = f & \text{in } \mathbb{R} \times \Omega, \\ \operatorname{div} u = 0 & \text{in } \mathbb{R} \times \Omega, \\ u = 0 & \text{auf } \mathbb{R} \times \partial\Omega \end{cases}$$

(LNSG)

für $\lambda > 0$. Wir beweisen neue Resultate zur Wohlgestelltheit, sowohl für stationäre als auch für zeitperiodische starke Lösungen zu (LNSG), in einem Außenraum $\Omega \subset \mathbb{R}^n$, $n \geq 2$, welche dann genutzt werden, um Existenz von stationären und zeitperiodischen Lösungen des zugehörigen nichtlinearen Systems unter geeigneten Kleinheitsbedingungen an die Daten zu zeigen. Während Lösungen zu diesen Außenraumproblemen üblicherweise im Rahmen von homogenen Sobolev-Räumen bestimmt werden, ist die Neuheit des vorgestellten Zugangs, dass er einen Rahmen liefert, in dem Lösungen in vollen Sobolev-Räumen nachgewiesen werden.

In Kapitel 4 untersuchen wir Problem (NSG) im dreidimensionalen Außenraum Ω unter geeigneten Annahmen an die Daten. Das zugehörige lineare Problem wird in Räumen absolut konvergenter Fourier-Reihen untersucht. Da das zugehörige Resolventenproblem in klassischen Sobolev-Räumen nicht wohlgestellt ist, leiten wir eine lineare Theorie in homogenen Sobolev-Räumen her. Diese neuen Resultate werden angewandt, um Existenz von zeitperiodischen Lösungen des nichtlinearen Systems (NSG) für „kleine" Daten zu zeigen.

In Kapitel 5 untersuchen wir das lineare Problem (LNSG) im Ganzraum $\Omega = \mathbb{R}^n$ für allgemeines $\lambda \in \mathbb{R}$, was wir als Problem auf der lokalkompakten abelschen Gruppe $\mathbb{T} \times \mathbb{R}^n$ formulieren. Wir führen zeitperiodische Fundamentallösungen für die Lösungen $(u, \mathfrak{p})$ dieser Gleichungen ein. Zusätzlich entwickeln wir das Konzept einer zeitperiodischen Fundamentallösung für die Wirbelstärke $\operatorname{curl} u$. Wir untersuchen Integrabilitätseigenschaften und zeigen punktweise Abschätzungen dieser Fundamentallösungen.

Das Thema von Kapitel 6 ist die Untersuchung asymptotischer Eigenschaften von zeitperiodischen Lösungen von (NSG) im Falle nichtverschwindender mittlerer Translationsgeschwindigkeit. Das Geschwindigkeitsfeld wird zerlegt in einen zeitunabhängigen Anteil und einen zeitperiodischen Restterm, und wir leiten separate punktweise Abschätzungen für diese beiden Anteile her. Hierdurch entdecken wir neue asymptotische Eigenschaften sowohl des Geschwindigkeitsfelds als auch der Wirbelstärke.

Contents

Contents

1 Introduction

1.1 The Navier–Stokes Equations

Introduced in the first half of the nineteenth century, the Navier–Stokes equations are still the most common way to model viscous incompressible fluid flows. They describe the flow inside a region $\Omega \subset \mathbb{R}^n$ during a time interval I by the equations

$$\begin{cases} \partial_t u + u \cdot \nabla u = f + \Delta u - \nabla \mathfrak{p} & \text{in } I \times \Omega, \\ \operatorname{div} u = 0 & \text{in } I \times \Omega, \end{cases} \tag{1.1}$$

where $u \colon I \times \Omega \to \mathbb{R}^n$ denotes the Eulerian velocity field and $\mathfrak{p} \colon I \times \Omega \to \mathbb{R}$ the pressure field of the fluid. The function $f \colon I \times \Omega \to \mathbb{R}^n$ denotes an external force. If Ω has a boundary, one usually adds appropriate boundary conditions for u to system (1.1), and if Ω is unbounded, also a boundary condition "at infinity", that is, a value for $\lim_{|x| \to \infty} u$, is included. The mathematical challenge consists of showing existence of solutions $(u, \mathfrak{p})$ to (1.1) and the investigation of their properties.

For simplicity, the viscosity and density constants are set to 1. While from a physical perspective only the case of a domain Ω in two or three dimensions seems relevant, mathematically, there is no reason not to formulate and study these equations in an n-dimensional domain $\Omega \subset \mathbb{R}^n$ for $n \geq 2$.

The mathematical examination of the Navier–Stokes equations is mostly carried out in one of three different frameworks. By including an initial condition $u(t_0, x) = u_0(x)$ at a time $t_0 \in I$ to problem (1.1), one obtains the corresponding *initial-value* problem. By restriction to time-independent quantities, one obtains the corresponding *steady-state* problem. Moreover, one can examine the *time-periodic* problem, where the involved functions exist on the whole time axis $I = \mathbb{R}$ and are periodic with respect to the time variable. The investigation of this latter case is subject of the present work.

Although being so different in their formulation, there exist many connections between these three types of problems. For instance, both steady-state and time-periodic solutions appear as equilibrium states of the evolution equation and thus as "limits" $t \to \infty$ of solutions to the initial-value problem. Another example are Hopf bifurcations, where the initially steady-state flow abruptly becomes time periodic. This phenomenon can be observed in physical experiments, for example, in the fluid flow past a body.

For the mathematical study of time-periodic solutions, it seems natural to treat them as specific solutions to the initial-value problem. Based on this idea, many mathematical concepts for the examination of time-periodic partial differential equations have been developed. Generally, time-periodic solutions found this way intrinsically have the same functional properties as solutions to the initial-value problem. In some settings this framework is not optimal to capture the functional properties of steady-state solutions. However, every steady-state solution is trivially also time periodic, which is why these should be included. For example, the steady-state flow past a body and the solution of the corresponding initial-value problem show very different behavior.

The study of a time-periodic problem as a particular initial-value problem has become very common in nowadays analysis. An overview of these methods is given below. Nevertheless, we use a different approach here that naturally includes the characteristics of the steady-state problem. The idea is based on the observation that a time-periodic function can be decomposed into the time mean over one period, which is time-independent and called *steady-state* part, and a second *purely periodic* part. The two parts can be investigated separately and, as is suggested by physical observations, possess different characteristic properties. While the steady-state part satisfies a time-independent problem and can be handled by classical methods, we investigate the purely periodic part with methods from the theory of Fourier analysis on groups. For this purpose, we model $\mathcal{T}$-time-periodic functions by means of the torus group $\mathbb{T} = \mathbb{R}/\mathcal{T}\mathbb{Z}$. This group naturally inherits a topology and a differentiability structure from $\mathbb{R}$, so that system (1.1) is equivalently formulated as

$$\begin{cases} \partial_t u + u \cdot \nabla u = f + \Delta u - \nabla \mathfrak{p} & \text{in } \mathbb{T} \times \Omega, \\ \operatorname{div} u = 0 & \text{in } \mathbb{T} \times \Omega. \end{cases} \tag{1.2}$$

The main advantage of this reformulation is that convolution and Fourier transform are available in this setting and the time periodicity is naturally included in the functional framework.

1.2 Time-Periodic Navier–Stokes Equations

While the foundation of the modern mathematical investigation of the Navier–Stokes equations can be dated back to the 1930s and the pioneering works of LERAY [78, 79], who established existence of weak solutions to the steady-state and the initial-value problem, the investigation of the time-periodic Navier–Stokes equations began much later, initiated by SERRIN [96] in the late 1950s.

It is remarkable that the study of general time-periodic partial differential equations started only some years earlier. By contemporary standards, the first mathematically rigorous contribution in this respect is due to PRODI [88] in the 1950s, who studied the time-periodic one-dimensional heat equation. Not much later, there appeared articles concerning the time-periodic wave equation [89, 34]. These papers can be seen as the basis on which more involved techniques were developed through the subsequent years.

Let us give a brief overview of the most common methods for the investigation of time-periodic differential equations and their first applications to the Navier–Stokes equations. To this end, consider the abstract evolution equation

$$\begin{cases} \partial_t u + Au = F(t, u) & \text{in } \mathbb{R}, \\ u(t + \mathcal{T}) = u(t), \end{cases} \tag{1.3}$$

where A is a linear (differential) operator on a Banach space X, and the (nonlinear) right-hand side F as well as the solution u are assumed to be $\mathcal{T}$-periodic for some prescribed period $\mathcal{T}$.

Probably, the most popular way to show existence of a solution to (1.3) is by means of fixed points of the so-called Poincaré operator. This operator maps a given initial value u_0 to the value $u(\mathcal{T})$ at time $\mathcal{T}$ of a solution *u* to the initial-value problem

$$\begin{cases} \partial_t u + Au = F(t, u) & \text{in } (0, \infty), \\ \quad u(0) = u_0. \end{cases} \tag{1.4}$$

This operator was introduced by POINCARÉ [86, 87], who considered it in the context of dynamical systems. Obviously, if u_0 is a fixed point of this mapping, then the solution *u* to (1.4) is $\mathcal{T}$-time-periodic and thereby also a solution to (1.3). In order to apply this procedure, the main challenge is to find a setting of Banach spaces such that the Poincaré operator is well defined and to ensure the existence of a fixed point. The first rigorous applications of this approach to investigate time-periodic differential equations directly in the infinite-dimensional framework go back to BROWDER [8], KRASNOSEL'SKIĬ [70] and KOLESOV [65, 66, 67].

In order to circumvent the more difficult analysis in infinite-dimensional spaces, one can combine this approach with classical energy methods and construct time-periodic solutions via a Galerkin approximation. In this way, one reduces (1.3) to a problem in a finite-dimensional setting and thus to an ordinary differential equation. In this framework, it is much easier to ensure the existence of a fixed-point of the corresponding Poincaré operator. The idea to use a Galerkin method in the field of time-periodic Navier–Stokes equations goes back to YUDOVICH [103] and PRODI [90]. However, their papers do not contain rigorous mathematical proofs. While the first application of the Galerkin method in the framework of the Navier–Stokes equations is due to HOPF [60], who constructed solutions to the initial-value problem, PROUSE [91, 92] connected this idea with the concept of the Poincaré operator to obtain time-periodic solutions.

Classically, by techniques from nonlinear functional analysis, one can obtain solutions to the nonlinear problem (1.3) from a suitable theory of the linear case where $F(t, u) = F(t)$, which can be investigated by means of the Poincaré operator as well. A different and more explicit approach is based on a specific representation formula for the solution. When the linear operator *A* is generator of a semi-group with suitable properties, one can establish the solution formula

$$u(t) = \int_{-\infty}^{t} e^{-(t-\tau)A} F(\tau) \, d\tau. \tag{1.5}$$

One readily verifies that *u* is time periodic of the same period as *F* if *u* is formally given by (1.5). Similarly to the approach based on the Poincaré operator, the difficulty is to find a suitable functional framework such that this representation formula is well defined. PRODI [88] was the first to investigate time-periodic partial differential equations by this method. In the study of time-periodic Navier–Stokes equations, the solution formula (1.5) was employed by KOZONO and NAKAO [69] and YAMAZAKI [102].

A different and more implicit method is to show that a solution u to the initial-value problem (1.4) tends to a periodic orbit as $t \to \infty$. Then the sequence (u_n) defined by $u_n(t) := u(t + n\mathcal{T})$ tends to a periodic solution to (1.3) as $n \to \infty$. This idea was first applied for general partial differential equations by FICKEN and FLEISHMAN [34], and SERRIN [96] initially proposed this method for the study of time-periodic Navier–Stokes equations. The first rigorous application in this field is due to KANIEL and SHINBROT [63], who showed existence of strong solutions in bounded domains for "small" data. TAKESHITA [99] extended their result to data of general "size" in the case of two dimensions.

The most classical way to study time-periodic problems is surely by a Fourier expansion in time, or equivalently, if we model time-periodic functions as functions on $\mathbb{T}$, by means of the Fourier transform on $\mathbb{T}$. In the linear case, the Fourier modes of the solution satisfy associated resolvent problems, which are partial differential equations only in spatial variables. This method has been applied by many researchers; see [89, 13, 93, 94, 58, 7] for example. The main difficulty is to transfer the *a priori* estimates, which are crucial to solve the original nonlinear problem, from the resolvent problems to the linearized time-periodic problem. Typically, this is only possible in two situations. On the one hand, one can use Plancherel's theorem to conclude estimates of the time-periodic solution in the space $L^2(\mathbb{T}; X)$, which is only possible if the underlying space X is a Hilbert space. On the other hand, one can establish estimates within a space of absolutely convergent Fourier series $A(\mathbb{T}; X)$, where all functions, in particular, the given data, are intrinsically continuous in time. One advantage of such a framework is that one can capture properties of each single Fourier mode specifically, which is why we employ this method in Chapter 4, where a time-periodic problem is examined in the spaces of the type $A(\mathbb{T}; X)$.

A recent approach to overcome the restriction to this kind of spaces is due to EITER, KYED and SHIBATA [23]. As uniform boundedness of the resolvent in the underlying space X is not sufficient to obtain *a priori* estimates for the time-periodic problem in the general Banach spaces $L^p(\mathbb{T}; X)$, they use the more restrictive notion of $\mathcal{R}$-boundedness which, in the end, is sufficient in this respect. In the last years, the concept of

$\mathcal{R}$-boundedness has become popular in the study of initial-value problems, and the article aims at directly transferring these results to the context of time-periodic problems. This approach can also be seen as an extension of the vector-valued multiplier theorems due to WEIS [101] and ARENDT and BU [4].

Another approach recently developed by GALDI [43, 44] and KYED [72, 73, 74] relies on a suitable decomposition of the time-periodic solution into two independent parts. The *steady-state* part is a solution to the corresponding steady-state problem, which can be treated by classical methods. The *purely periodic* part can be investigated separately and thus in a completely different functional framework. As we observe in Chapter 3, the investigation of the purely periodic part can usually be achieved in a functional framework with simpler structure than that of the steady-state part. Moreover, the steady-state and purely periodic parts usually show different decay and integrability properties. This characteristic also appears in our analysis in Chapter 5 and Chapter 6. In this respect, the decomposition allows to appropriately capture the physical properties of the investigated system. Note that this method has also found application in other fields besides fluid dynamics. For example, it was employed by KYED and CELIK [11, 12] in order to study damping effects in different nonlinear wave equations, and by IBRAHIM, LEMARIÉ-RIEUSSET and MASMOUDI [61] in the investigation of time-periodic solutions to the Navier–Stokes–Maxwell equations.

1.3 Flow Around a Moving Body

The focus of the present work lies in the investigation of time-periodic flow around a rigid body $\mathcal{B}$ that performs a prescribed motion through a Navier–Stokes liquid. In this case the region of flow depends on time, that is, $\Omega = \Omega(t)$. In view of (1.1), the Navier–Stokes flow is then described by the equations

$$\begin{cases} \partial_t u + u \cdot \nabla u = f + \Delta u - \nabla \mathfrak{p} & \text{in } \bigcup_{t \in I} \{t\} \times \Omega(t), \\ \text{div } u = 0 & \text{in } \bigcup_{t \in I} \{t\} \times \Omega(t), \\ u = U_{\mathcal{B}} & \text{on } \bigcup_{t \in I} \{t\} \times \partial\Omega(t), \\ \lim_{|x| \to \infty} u(t, x) = 0 & \text{for } t \in I. \end{cases} \qquad (1.6)$$

Here $U_\mathcal{B}$ describes the velocity of the boundary of the body $\mathcal{B}$. Hence, condition $(1.6)_3$ means that the fluid particles adhere to the body $\mathcal{B}$ at the boundary. Moreover, $(1.6)_4$ means that the fluid flow is at rest "at infinity".

Observe that it is also possible to prescribe a non-vanishing (time-dependent) velocity u_∞ at infinity by replacing equation $(1.6)_4$ with the condition $\lim_{|x|\to\infty} u(t,x) = u_\infty(t)$. However, by a simply change of coordinates, this merely corresponds to an additional translational velocity of the body $\mathcal{B}$, so that this case is included in the above description.

Instead of a description of the flow in an inertial frame as in (1.6), it is often convenient to change coordinates and to describe the problem in a frame attached to the body $\mathcal{B}$. In the three-dimensional case, one can express (1.6) in the form

$$\begin{cases} \partial_t u + \eta \wedge u - \eta \wedge x \cdot \nabla u - \xi \cdot \nabla u + u \cdot \nabla u = f + \Delta u - \nabla \mathfrak{p} & \text{in } \mathbb{R} \times \Omega, \\ \operatorname{div} u = 0 & \text{in } \mathbb{R} \times \Omega, \\ u = \xi + \eta \wedge x & \text{on } \mathbb{R} \times \partial\Omega, \\ \lim_{|x|\to\infty} u(t,x) = 0 & \text{for } t \in \mathbb{R}, \end{cases}$$

$$(1.7)$$

where Ω is the exterior of the body $\mathcal{B}$; see [41] for example. Here ξ and η correspond to (possibly time-dependent) translational and angular velocity of the rigid motion of the body $\mathcal{B}$. Since we study time-periodic solutions, we have replaced the time axis I by $\mathbb{R}$. Both physical and mathematical observations show that properties of the fluid flow strongly depend on ξ and η, which is why the investigation is usually carried out for different cases separately. Note that in the absence of rotation, the parameter η vanishes, and (1.7) is a proper description of the fluid flow in any dimension $n \geq 2$.

Observe that the domain in (1.7) is an *unbounded* exterior domain $\Omega \subset \mathbb{R}^n$. While the results listed above mostly concerned the study of the flow in bounded domains, a proof of the existence of strong time-periodic solutions to the Navier–Stokes equations in unbounded domains was achieved in the 1990s by MAREMONTI [81, 82] and extended by MAREMONTI and PADULA [83] in the case $\xi = \eta \equiv 0$. Their result was complemented by a different proof by KOZONO and NAKAO [69] based on the representation formula (1.5), but they could not treat the case of a three-dimensional exterior domain. A short time later this case was covered by YAMAZAKI [102], who could show existence in the framework of so-called mild solutions in a framework of weak L^p spaces, and subsequently by GALDI and SOHR [38], who showed existence of strong time-periodic solutions in spatially weighted spaces.

While all these works only examine the case $\xi = \eta \equiv 0$, that is, the time-periodic flow around a body at rest, the case of a moving body has attracted less attention. The first investigation in this respect is due to GALDI and SILVESTRE [53], who showed existence of weak solutions for time-periodic data ξ, η and f. They further developed these results to the case of a freely moving body in [54]. These results were established in a Hilbert-space framework based on energy methods. One drawback of this framework is that it does not allow to capture asymptotic properties of the velocity field. Therefore, an investigation of the problem in an L^p setting for $p \neq 2$ is required.

From a physical point of view, it is reasonable to introduce

$$\lambda := \left| \frac{1}{\mathcal{T}} \int_0^{\mathcal{T}} \xi \, \mathrm{d}t \right|, \tag{1.8}$$

which is the modulus of the time mean of the translational velocity ξ, and to distinguish the cases $\lambda = 0$ and $\lambda \neq 0$. If $\lambda = 0$, then the body $\mathcal{B}$ oscillates around a prescribed point in space, and if $\lambda \neq 0$, it performs a proper translation. The parameter λ has significant impact on the physical properties of the described flow. One difference can be observed in the asymptotic behavior of the flow. For $\lambda \neq 0$ there is a wake region "behind" the body, which does not exist for $\lambda = 0$. In this respect, it seems reasonable to also distinguish these cases in the mathematical investigation. Since the case $\lambda \neq 0$ is equivalent to the configuration where the body oscillates at a prescribed place and the velocity "at infinity" is prescribed by $\lim_{|x| \to \infty} u(t, x) = -\lambda e_1$, the case $\lambda \neq 0$ is also called the time-periodic flow *past* a body, in contrast to the case $\lambda = 0$, the flow *around* a body.

The case of a flow past a non-oscillating and non-rotating body, that is, the case $|\xi| \equiv \lambda \neq 0$ and $\eta = 0$, was examined by KYED [73, 74], who showed existence of time-periodic solutions in a framework of L^p spaces. The same case was treated by by GALDI [43, 44] in two dimensions. Both used the method of decomposing the time-periodic solution into a steady-state and a purely periodic part. Furthermore, GALDI and KYED [50] showed existence of time-periodic solutions in three-dimensional exterior domains for the problem of a flow past an oscillating body, that is, the case of time-periodic ξ with $\lambda \neq 0$ and $\eta = 0$. Moreover, GALDI [47] recently proved existence of solutions to the problem of time-periodic flow around an oscillating body, that is, the case of time-periodic ξ with $\lambda = 0$, in spatially weighted function spaces. However, these papers always consider the case $\eta = 0$ of a non-rotating body. Following a completely different approach, GEISSERT, HIEBER and NGUYEN [55] established the existence of mild solutions to the problem of time-periodic flow past a rotating body, where both ξ and η are time-independent and parallel.

1.3.1 Flow Past a Non-rotating Body

Chapter 3 is concerned with the case of time-periodic flow past a body that performs a purely translational motion, that is, the case of constant translational velocity $\xi \neq 0$ and vanishing angular velocity $\eta = 0$. By choosing the coordinate system appropriately, we may assume $\xi = \lambda\,e_1$ for some $\lambda > 0$. Our analysis of time-periodic solutions to (1.6) is mainly based on the investigation of the associated linear system. As explained above, for a prescribed time period $\mathcal{T} > 0$ we model $\mathcal{T}$-time-periodic functions as functions on the torus group $\mathbb{T} = \mathbb{R}/\mathcal{T}\mathbb{Z}$. Then the linearization of (1.7) is given by

$$
\begin{cases}
\partial_t u - \Delta u - \lambda\partial_1 u + \nabla\mathfrak{p} = f & \text{in } \mathbb{T}\times\Omega, \\
\operatorname{div} u = 0 & \text{in } \mathbb{T}\times\Omega, \\
u = 0 & \text{on } \mathbb{T}\times\partial\Omega
\end{cases}
\tag{1.9}
$$

for a right-hand side $f\colon\mathbb{T}\times\Omega\to\mathbb{R}^n$. Observe that we omitted the boundary condition $\lim_{|x|\to\infty} u(t,x) = 0$ "at infinity" in this formulation. Since we search for solutions $(u,\mathfrak{p})$ such that $u(t,\cdot)\in L^q(\Omega)$ for some $q\in(1,\infty)$, the velocity field u is intrinsically subject to this condition in a generalized sense. We call (1.9) the time-periodic Stokes problem if $\lambda = 0$, and the time-periodic Oseen problem if $\lambda\neq 0$. We follow the approach by GALDI and KYED and decompose problem (1.9) into two separate problems, which can be examined independently. To this end, we introduce the projection operators

$$
\mathcal{P}f := \frac{1}{\mathcal{T}}\int_0^{\mathcal{T}} f\,dx, \qquad \mathcal{P}_\perp = \operatorname{Id}-\mathcal{P}.
$$

Then $\mathcal{P}f$ is the time mean of the function f and a *time-independent* function, and $f = \mathcal{P}f + \mathcal{P}_\perp f$ is decomposed into *steady-state part* and $\mathcal{P}f$ and *purely periodic part* of $\mathcal{P}_\perp f$. By means of these projectors, we decompose all functions in (1.9) and define

$$
v := \mathcal{P}u, \quad w := \mathcal{P}_\perp u, \quad p := \mathcal{P}\mathfrak{p}, \quad \mathfrak{q} := \mathcal{P}_\perp\mathfrak{p}, \quad g := \mathcal{P}f, \quad h := \mathcal{P}_\perp f.
$$

In this way, (1.9) is separated into two systems, namely

$$\begin{cases} -\Delta v - \lambda\partial_1 v + \nabla p = g & \text{in } \Omega, \\ \operatorname{div} v = 0 & \text{in } \Omega, \\ v = 0 & \text{on } \partial\Omega. \end{cases} \qquad (1.10)$$

and

$$\begin{cases} \partial_t w - \Delta w - \lambda\partial_1 w + \nabla\mathfrak{q} = h & \text{in } \mathbb{R}\times\Omega, \\ \operatorname{div} w = 0 & \text{in } \mathbb{R}\times\Omega, \\ w = 0 & \text{on } \mathbb{R}\times\partial\Omega. \end{cases} \qquad (1.11)$$

In analogy to before, if $\lambda = 0$, we call these equations the steady-state and purely periodic Stokes problems, and if $\lambda \neq 0$, we call them the steady-state and purely periodic Oseen problems. Though looking quite the same at first glance, the main difference between (1.9) and (1.11) is that in the latter there appear only purely periodic functions. The advantage of this decomposition is that one can now search for functional frameworks that render (1.10) and (1.11) well posed independently of each other. By combination, one then obtains a comprehensive solution theory for the original time-periodic problem (1.9).

As mentioned above, the fluid flow shows different physical properties depending on the parameter λ, which is called Reynolds number. Whether $\lambda = 0$ or $\lambda \neq 0$, also has impact on the analysis of the linear problem (1.9), and the functional analytic properties differ significantly. However, as a further investigation shows, this discrepancy is only due to the steady-state part and one can establish a solution theory for the purely periodic problem (1.11) that is uniform in λ. In particular, as in the case of steady-state flow, the linear solution theory in an L^q framework is sufficient for the treatment of the nonlinear problem (1.1) by a fixed-point argument; see [73]. In contrast, for $\lambda = 0$ one obtains different function spaces and the nonlinear problem cannot be treated in this manner. In the present work, we therefore focus on the case of non-vanishing Reynolds number $\lambda \neq 0$, that is, the case of flow *past* a body.

As explained before, our approach to the time-periodic linear problem (1.9) requires an appropriate theory for strong solutions to its steady-state counterpart (1.10), which can be regarded as a special case. The first fundamental contribution in this regard is due to GALDI [40], who established frameworks of well-posedness for weak and strong solutions to the steady-state problem (1.10). An overview and further articles can be found in [42, Chapter VII], the more recent paper [2] and the references therein. As the investigation of (1.10) shows, for general $g \in L^q(\Omega)$ with suitable $q \in (1, \infty)$ the corresponding velocity-field solution merely satisfies

$$\nabla^2 v \in L^q(\Omega), \qquad \nabla v \in L^{s_1}(\Omega), \qquad v \in L^{s_2}(\Omega)$$

for different values $q < s_1 < s_2$, but v is not an element of the *full* Sobolev space $W^{2,q}(\Omega)$. In contrast, the velocity-field solution w to the purely periodic linear problem (1.11) shows a completely different behavior and belongs to the full Sobolev space $W^{2,q}(\Omega)$ with respect to the spatial variable. Combining these two results, one concludes well-posedness of the complete time-periodic problem (1.9) in a functional framework that is suitable to show existence of a solution to the nonlinear problem (1.7) in three-dimensional exterior domains; see [50].

In Chapter 3 we derive a different solution theory for the time-periodic problem (1.9). While the purely periodic velocity field w belongs to the same space as in [50], the novelty of the presented approach is the construction of a framework that ensures the steady-state solution v to belong to the full Sobolev space $W^{2,q}(\Omega)$ as well. In consequence, we also obtain a velocity-field solution u to the time-periodic system (1.9) that is an element of $W^{2,q}(\Omega)$ with respect to space. From this result, we finally conclude existence of steady-state and time-periodic solutions to the Navier–Stokes problem (1.7) with $\xi = \lambda e_1 \neq 0$ and $\omega \neq 0$ such that the velocity field belongs to this full Sobolev space. The results of Chapter 3 were published by EITER and GALDI [19].

This investigation in Chapter 3 is motivated by recent research on time-periodic bifurcations by GALDI [46, 45], who studied Hopf bifurcations in the context of the flow past a body, which were established in a framework where the steady-state part of the solution merely belongs to homogeneous Sobolev spaces as described above. We assume that the results presented here allow to study these Hopf bifurcation as well as secondary bifurcations in a framework of full Sobolev spaces and to make them accessible for techniques typically used in bounded domains. The mathematical proof of secondary bifurcation for fluid flow problems in exterior domains is still an open problem to date.

Observe that the occurrence of Hopf bifurcations is one of the most natural appearances of time-periodic flows in physics. While GALDI studied bifurcations that occur in the flow past a body, another example is the spontaneous oscillation of a falling drop when its falling velocity exceeds a specific value. The mathematical examination of this phenomenon has recently been initiated by EITER, KYED and SHIBATA, who established an appropriate framework for the existence of steady-state solutions [24] and investigated existence of solutions to the corresponding time-periodic problem in bounded domains [25].

1.3.2 Flow Past a Rotating Body

The subject of Chapter 4 is the time-periodic flow past a rotating body in the three-dimensional whole space. More precisely, we consider system (1.6) in the case of a time-periodic translation velocity ξ with time average $\lambda \neq 0$ and a non-vanishing angular velocity η. We assume that the translation velocity ξ and the external force f are time periodic with coincident period $\mathcal{T} > 0$, that is,

$$\xi(t+\mathcal{T}) = \xi(t), \qquad f(t+\mathcal{T},x) = f(t,x).$$

Moreover, we assume that the axis of translation does not vary over time and coincides with the rotational axis, and that, without loss of generality, both are directed along the x_1-axis. This means

$$\xi(t) = \alpha(t)\,e_1, \qquad \eta = \omega\,e_1 \tag{1.12}$$

for some prescribed $\mathcal{T}$-periodic function $\alpha \colon \mathbb{R} \to \mathbb{R}$ and a constant $\omega \in \mathbb{R} \setminus \{0\}$.

We further assume that the body performs a proper translation such that after one period its center of mass has changed its location. Expressed differently, we assume the mean translational velocity of the body over one time period to be non-zero:

$$\lambda := \frac{1}{\mathcal{T}} \int\limits_0^{\mathcal{T}} \alpha(t)\,dt \neq 0. \tag{1.13}$$

Observe that this parameter λ coincides with λ defined in (1.8) provided $\lambda > 0$, which we may assume without loss of generality. In view of these specifications, the linearization of (1.7) is given by

$$\begin{cases} \partial_t u + \omega(e_1 \wedge u - e_1 \wedge x \cdot \nabla u) - \Delta u - \lambda \partial_1 u + \nabla \mathfrak{p} = f & \text{in } \mathbb{T} \times \Omega, \\ \operatorname{div} u = 0 & \text{in } \mathbb{T} \times \Omega, \\ u = 0 & \text{on } \mathbb{T} \times \partial\Omega. \end{cases} \tag{1.14}$$

Here the parameter λ plays the same role as discussed above, and physical and mathematical properties heavily depend on whether or not (1.13) is satisfied. In particular, the case $\lambda = 0$ would require different techniques and a different functional framework, which is why it is not considered here. In the case $\lambda \neq 0$, problem (1.14) is called the time-periodic generalized Oseen problem.

Observe that, since we consider the case $\omega \neq 0$ of a proper rotation, the term $\omega\, e_1 \wedge x \cdot \nabla$ appearing in (1.14) does not vanish and is a differential operator with *unbounded* coefficient. Therefore, even for "small" η, problem (1.14) cannot be treated as a lower-order perturbation of the time-periodic Oseen problem (1.9). Moreover, we find that the analysis of the corresponding resolvent problem requires a completely different functional setting. In order to find a framework that renders the time-periodic generalized Oseen problem (1.14) well posed, we make use of the fact that the term $\eta \wedge x \cdot \nabla$ stems from a change of coordinates into a frame attached to the body and can be dealt with by the reversed transform. This idea was also employed by GALDI and KYED [49, 48] to investigate the the steady-state problem corresponding to (1.7). However, in the time-periodic framework, this method only yields suitable estimates when the change of coordinates maintains the time periodicity of the involved functions. Observe that this is the case if the angular velocity ω of the rotation of the body coincides with an integer multiple of the angular frequency $2\pi/\mathcal{T}$ of the time-periodic data. For the sake of simplicity, we assume

$$\omega = 2\pi/\mathcal{T}. \tag{1.15}$$

This assumption means that after one time period the rigid body completed one full revolution. Regarding the body rotation as a second time-periodic external forcing mechanism, one may interpret (1.15) as the condition that this mechanism has to be compatible to the time-periodic data. In the end, we show existence of a time-periodic solution to (1.7) under the assumptions (1.12), (1.13) and (1.15). These results were published in [22].

1.3.3 Asymptotic Behavior

The second main topic of the present work is the analysis of the asymptotic behavior of a time-periodic fluid flow surrounding a moving obstacle, which is governed by the Navier–Stokes equations (1.6). From information on the asymptotic properties of the flow one can directly conclude physical properties. For example, the anisotropic nature of the derived estimates hints at the occurrence of a wake region "behind" the body. Moreover, these estimates show that the purely periodic part of both the velocity and the vorticity field decay faster than the corresponding steady-state part, so that the characteristics of the far field of a time-periodic flow are the same as observed for steady-state flows.

One classical approach to study asymptotic properties is by means of a fundamental solution. While for the steady-state linearized Navier–Stokes problem (1.10) fundamental solutions are known for many decades and go back to LORENTZ [80] and OSEEN [85], corresponding results for the time-periodic linearized Navier–Stokes problem

$$\begin{cases} \partial_t u - \Delta u - \lambda \partial_1 u + \nabla \mathfrak{p} = f & \text{in } \mathbb{T} \times \mathbb{R}^n, \\ \operatorname{div} u = 0 & \text{in } \mathbb{T} \times \mathbb{R}^n, \end{cases} \tag{1.16}$$

were introduced by KYED [76] and GALDI and KYED [51] very recently in dimension $n = 3$. The subject of Chapter 5 is an extension of their results to general dimension $n \geq 2$, which was published by EITER and KYED [21]. Moreover, we introduce a fundamental solution associated to the vorticity field of time-periodic linearized Navier–Stokes flow (1.16) in $n = 3$ dimensions.

One notable property of these time-periodic fundamental solutions is that they are also subject to the decomposition explained above and can be identified as the sum of a steady-state part, which coincides with the fundamental solution to the respective steady-state problem, and a purely periodic part, which possesses better properties in terms of decay and integrability. Since the steady-state problem is a special case of the time-periodic problem, it is no surprise to recover the steady-state fundamental solution in the time-periodic framework.

Because we formulated the time-periodic whole-space problem (1.16) as a problem in the locally compact abelian group $G = \mathbb{T} \times \mathbb{R}^n$, where a Fourier transform is available, the second, purely periodic part can be identified in terms of a Fourier multiplier on this group. This enables us to investigate time-periodic fundamental solutions by means of Fourier analytic methods in the group G. In particular, we establish integrability properties and pointwise estimates, which are subsequently employed to study asymptotic properties of time-periodic flows.

As in the case of a steady-state flow, the fundamental solution associated to the velocity u in (1.16) shows a polynomial decay, and the decay rate of the fundamental solution associated to the vorticity $\operatorname{curl} u$ is of exponential type. Remarkably, while in the case $\lambda \neq 0$ the decay rate of the steady-state part is anisotropic and decays faster outside a wake region, the purely periodic part behaves differently and decays homogeneous in space.

The properties established for the fundamental solutions allow to ana-lyze the asymptotic structure of time-periodic flow. Recall that the case of a steady-state Navier–Stokes flow, described by

$$\begin{cases} \eta \wedge u - \eta \wedge x \cdot \nabla u - \xi \cdot \nabla u + u \cdot \nabla u = f + \Delta u - \nabla \mathfrak{p} & \text{in } \Omega, \\ \qquad\qquad\qquad \text{div } u = 0 & \text{in } \Omega, \\ \qquad\qquad\qquad u = \xi + \eta \wedge x & \text{on } \partial\Omega, \\ \qquad\quad \lim_{|x|\to\infty} u(x) = 0, \end{cases} \tag{1.17}$$

can be regarded as a special case of a time-periodic flow. The mathemati-cal analysis of the asymptotic behavior of solutions to (1.17) can be dated back to FINN [36], who showed that the asymptotic profile of the velocity field u of the flow past a (non-rotating) body, which means $\xi \neq 0$ and $\eta = 0$, is dominated by that of the steady-state Oseen fundamental solu-tion. Later BABENKO [5] extended this result to weak solutions. However, his proof had gaps and was finally completed by GALDI [39]. Later, KYED [75] obtained an analogous result in the case of the flow past a rotating body. The asymptotic structure of the associated vorticity field curl u was identified by CLARK [14] and BABENKO and VASIL'EV [6] in the case of a Navier–Stokes flow past a non-rotating body. Their result was recently extended to the case of a rotating body by DEURING and GALDI [17].

In the case $\xi = \eta = 0$, that is, the case of the flow around a body at rest, an asymptotic expansion for the velocity field was established by KOROLEV and ŠVERAK [68], under the assumption of "small" data. They identified the corresponding leading term as a so-called Landau solution of the Navier–Stokes equations. If the body performs only a rotation but no translation, that is, if $\xi = 0$ and $\eta \neq 0$, FARWIG and HISHIDA [29] and FARWIG, GALDI and KYED [28] established similar expansions with the same leading terms.

Concerning the time-periodic case, which is described by time-periodic solutions to (1.7), KANG, MIURA and TSAI [62] recently derived an asymp-totic expansion for the velocity field u for flow around a body at rest, that is, in the case $\xi = \eta = 0$. For the flow past a non-rotating body, that is, in the case of constant translational velocity $\xi \neq 0$ and vanishing angu-lar velocity $\eta = 0$, an asymptotic expansion was established by GALDI and KYED [51]. In both results the asymptotic profile of the time-periodic flow coincides with that of the corresponding steady-state problem described above.

In Chapter 6 we further investigate the case of flow past a body. More precisely, in the final result, the mean translational velocity ξ is allowed to be time dependent with non-vanishing time mean λ defined in (1.8). Beginning with the asymptotic expansion established in [51], we employ the time-periodic fundamental solutions from Chapter 5 to derive pointwise estimates of the velocity field u and its gradient ∇u of a solution $(u, \mathfrak{p})$ to (1.7). Subsequently, we examine the corresponding vorticity field $\operatorname{curl} u$ by means of an integral representation via the associated fundamental solution.

One of our main observations in Chapter 5 is that the steady-state parts of all investigated time-periodic fundamental solutions decay slower than the associated purely periodic parts. Our findings in Chapter 6 show that the velocity and vorticity corresponding to a time-periodic Navier–Stokes flow have similar properties, and we observe that the steady-state parts of both the velocity and vorticity fields decay slower than the respective purely periodic parts.

2 Preliminaries

In this chapter, we prepare the notation used throughout the present thesis as well as some preliminary results. After introducing the basic notation and function spaces, we present some results from harmonic analysis on groups. For a more detailed introduction to this topic, we refer to [95]. Subsequently, we introduce the notation for Sobolev spaces together with frequently used inequalities. Finally, we collect some preliminary results from mathematical fluid dynamics. Most of the presented results can be found in [42].

2.1 Basic Notation

In this section we prepare the basic notation concerning number sets, vector analysis and function spaces.

2.1.1 Sets of Numbers

The symbol $\mathbb{N}$ denotes the set of natural numbers, that is, the set of positive integers. We set $\mathbb{N}_0 := \mathbb{N} \cup \{0\}$, and let $\mathbb{Z}$, $\mathbb{R}$ and $\mathbb{C}$ denote the sets of integers and real and complex numbers, respectively. Usually, elements in $\mathbb{R} \times \mathbb{R}^n$ are denoted by (t, x), and consist of a time variable t and a spatial variable x. For the sign of a real number $x \in \mathbb{R}$ we write $\mathrm{sgn}(x)$, and the argument of a complex number $z \in \mathbb{C} \setminus \{0\}$ is denoted by $\arg z \in (-\pi, \pi]$.

For $n \in \mathbb{N}$ let $\alpha = (\alpha_1, \ldots, \alpha_n) \in \mathbb{N}_0^n$ and $\beta = (\beta_1, \ldots, \beta_n) \in \mathbb{N}_0^n$ be multi-indices. Then we set $|\alpha| = \alpha_1 + \cdots + \alpha_n$. The notation $\alpha \leq \beta$ means $\alpha_j \leq \beta_j$ for all $j = 1, \ldots, n$.

When considering a number $a > c$ for $c \in \mathbb{R}$, we implicitly assume that a is real if not indicated otherwise. However, there is one exception of this rule: If a is replaced by the letter n, that is, when considering $n > c$ for some $c \in \mathbb{R}$, we implicitly assume that $n \in \mathbb{N}$.

The set $\{e_1, \ldots e_n\}$ denotes the standard basis of $\mathbb{R}^n$ and $\mathbb{C}^n$. Then $e_j \cdot e_k = \delta_{jk}$, where δ_{jk} denotes the Kronecker delta. Let $x = (x_1, \ldots, x_n)$ and $y = (y_1, \ldots, y_n)$ be elements of $\mathbb{R}^n$ or $\mathbb{C}^n$. We denote the Euclidean norm of x by

$$|x| := \sqrt{|x_1|^2 + \cdots + |x_n|^2},$$

and for the (real) Euclidean scalar product we write $x \cdot y := x_j y_j$. Here and in the following we use the Einstein summation convention, that is, in products we implicitly sum over repeated indices from 1 to n. Similarly, for matrices $A = (A_{jk})$, $B = (B_{jk}) \in \mathbb{C}^{n \times n}$ the scalar product is denoted by $A : B := A_{jk} B_{jk}$. Moreover, the tensor product $x \otimes y \in \mathbb{C}^{n \times n}$ of x and y is given by $(x \otimes y)_{jk} = x_j y_k$. The vector product $x \wedge y \in \mathbb{C}^3$ of $x, y \in \mathbb{C}^3$ is defined by $(x \wedge y)_j = \varepsilon_{jk\ell} x_k y_\ell$, where $\varepsilon_{jk\ell}$ is the Levi-Civita symbol defined by

$$\varepsilon_{jk\ell} := \begin{cases} 1 & \text{if } (j, k, \ell) \text{ is an even permutation of } (1, 2, 3), \\ -1 & \text{if } (j, k, \ell) \text{ is an odd permutation of } (1, 2, 3), \\ 0 & \text{otherwise.} \end{cases}$$

Moreover, if $z \in \mathbb{C}^3$ is a third vector, we set $x \wedge y \cdot z := (x \wedge y) \cdot z$.

For $R > 0$ and $x \in \mathbb{R}^n$ we let $\mathrm{B}_R(x) \coloneqq \{y \in \mathbb{R}^n \mid |x - y| < R\}$ denote the ball of radius R centered at x, and $\mathrm{B}^R(x) \coloneqq \{y \in \mathbb{R}^n \mid |x - y| > R\}$ is the interior of its complement. For $x = 0$ we simply write $\mathrm{B}_R \coloneqq \mathrm{B}_R(0)$ and $\mathrm{B}^R \coloneqq \mathrm{B}^R(0)$. Moreover, ω_n denotes the surface area of unit sphere in $\mathbb{R}^n$. For $\Omega \subset \mathbb{R}^n$ we define $\Omega_R \coloneqq \Omega \cap \mathrm{B}_R$ and $\Omega^R \coloneqq \Omega \cap \mathrm{B}^R$.

A set $\Omega \subset \mathbb{R}^n$ is called a domain if it is a non-empty, open, connected subset of $\mathbb{R}^n$. We call Ω a bounded domain if it is contained in a ball B_R for some radius $R > 0$, and an exterior domain if it is the complement of the closure K of a bounded domain in $\mathbb{R}^n$. In the latter case, $\delta(\Omega^c)$ denotes the diameter of K. Without loss of generality, we make the technical assumption that 0 belongs to the interior of K.

If the (compact) boundary of Ω can locally be represented as the graph of a Lipschitz continuous function, we say that Ω has Lipschitz boundary or that it is a Lipschitz domain. If it can locally be represented by the graph of a function that is k-times continuously differentiable, $k \in \mathbb{N}$, we call Ω a domain of class C^k, a domain with C^k-boundary or a C^k-domain.

Spatial derivatives of a sufficiently regular function u on a domain $\Omega \subset \mathbb{R}^n$ are denoted by $\partial_j u = \partial_{x_j} u$ for $j = 1, \dots, n$, and $\nabla u \coloneqq (\partial_1 u, \dots, \partial_n u)$ denotes the gradient of u. We set $\mathrm{D}^\alpha u \coloneqq \mathrm{D}_x^\alpha u \coloneqq \partial_1^{\alpha_1} \dots \partial_n^{\alpha_n} u$ for $\alpha \in \mathbb{N}_0^n$, and the symbol $\nabla^k u = (\mathrm{D}^\alpha u \mid \alpha \in \mathbb{N}_0^n, |\alpha| = k)$ denotes the formal collection of all derivatives of order $k \in \mathbb{N}$. Moreover, $\Delta u = \partial_j \partial_j u$ defines the Laplace operator. If $u = (u_1, \dots, u_n)$ is $\mathbb{R}^n$-valued, these differential operators act on u componentwise. Moreover, $\mathrm{div}\, u = \partial_j u_j$ denotes the divergence of u. If v is another $\mathbb{R}^n$-valued function on Ω, then $u \cdot \nabla v$ is the vector-valued function defined by $(u \cdot \nabla v)_j = u_k \partial_k v_j$. We further let $\mathrm{curl}\, u$ denote the curl or rotation of an $\mathbb{R}^3$-valued vector field u, that is, $(\mathrm{curl}\, u)_j = \varepsilon_{jk\ell} \partial_k u_\ell$. For a second-order tensor field $T \colon \mathbb{R}^n \to \mathbb{R}^{n \times n}$ the vector field $\mathrm{div}\, T$ is defined by $(\mathrm{div}\, T)_j \coloneqq \partial_k T_{jk}$. Unless indicated otherwise, these operators always act on the spatial variables x, also in the case when u further depends on time. In this case, $\partial_t u$ denotes its time derivative.

We use capital letters to denote global constants, which are numbered consecutively throughout the complete thesis, and we use small letters to denote local constants, which are numbered in the respective proof. In order to emphasize that a constant C depends on quantities $\alpha, \beta, \gamma, \dots$, we write $C = C(\alpha, \beta, \gamma, \dots)$.

2.1.2 Periodic Functions

Let X be any set and $u \colon \mathbb{R} \to X$ be a periodic function of period $\mathcal{T} > 0$, that is, $u(t + \mathcal{T}) = u(t)$ for all $t \in \mathbb{R}$. When u depends on further variables and is periodic considered as a function of the time variable t, we call u time periodic.

For a fixed period $\mathcal{T}$, we write $\mathbb{T} = \mathbb{R}/\mathcal{T}\mathbb{Z}$ for the corresponding torus group, which is naturally equipped with the associated quotient topology. Let $\pi\colon\mathbb{R} \to \mathbb{R}/\mathcal{T}\mathbb{Z}$, $t \mapsto [t]$ be the quotient map from $\mathbb{R}$ onto $\mathbb{R}/\mathcal{T}\mathbb{Z}$, and let $\Pi\colon\mathbb{T} \to [0,\mathcal{T})$ be the function that maps each coset $[t] \in \mathbb{T} = \mathbb{R}/\mathcal{T}\mathbb{Z}$ to its unique representative $t \in [0,\mathcal{T})$. For any $\mathcal{T}$-periodic function $u\colon\mathbb{R} \to X$, the composition $U = u \circ \Pi$ is a function $\mathbb{T} \to X$, and for any function $U\colon\mathbb{T} \to X$, the composition $u = U \circ \pi$ is a $\mathcal{T}$-periodic function $\mathbb{R} \to X$. In this way, we can identify $\mathcal{T}$-periodic functions on $\mathbb{R}$ with functions on $\mathbb{T} = \mathbb{R}/\mathcal{T}\mathbb{Z}$. In the following, we shall always do this tacitly and simply write $U = u$ in this case. In particular, also $t \in \mathbb{T}$ is referred to as the time variable.

Besides its topology, $\mathbb{T} = \mathbb{R}/\mathcal{T}\mathbb{Z}$ inherits also the differentiability structure of $\mathbb{R}$, and we say that a function $U\colon\mathbb{T} \to X$ is k-times (continuously) differentiable if $u = U \circ \pi$ is k-times (continuously) differentiable, and the (time) derivative is defined by $\partial_t U = \partial_t u \circ \pi$.

2.1.3 General Vector Spaces

For topological vector spaces X and Y the symbol $\mathcal{L}(X,Y)$ denotes the set of all continuous linear operators $X \to Y$, and we set $\mathcal{L}(X) := \mathcal{L}(X;X)$. We write I for the identity mapping in X. The (topological) dual space of X is denoted by X', and for the corresponding dual pairing of $x \in X$ and $\varphi \in X'$ we write $\langle\varphi, x\rangle = \varphi(x)$. If X is a semi-normed vector space, we denote its semi-norm by $\|\cdot\|_X$. For semi-normed vector spaces X and Y, the Cartesian product $X \times Y$ is usually equipped with the semi-norm

$$\|(x,y)\|_{X\times Y} := \|x\|_X + \|y\|_Y.$$

If $X, Y \subset Z$ for some vector space Z, then their intersection $X \cap Y$ is equipped with the semi-norm

$$\|z\|_{X\cap Y} := \|z\|_X + \|z\|_Y,$$

and if $X \cap Y = \{0\}$, then $X \oplus Y$ denotes their direct sum. In general, our notation does not distinguish between the semi-norm of a vector space and the semi-norm of the n-times Cartesian product X^n, and we simply write $\|\cdot\|_X$ for the semi-norm in both cases. Moreover, when the dimension is clear from the context, we occasionally omit the exponent n and simply write $x \in X$ instead of $x \in X^n$.

If X is a complex-valued vector space and $A \subset X$, then $\mathrm{span}_{\mathbb{C}} A$ denotes the linear hull of A in X.

2.1.4 Continuous and Integrable Functions

Let X and Y be topological spaces. The space of all continuous functions $X \to Y$ is denoted by $\mathrm{C}(X;Y)$.

For X an open subset of $\mathbb{R}^n$ or $\mathbb{T} \times \mathbb{R}^n$ and Y a normed vector space, the set $\mathrm{C}^k(X;Y)$ contains all k-times continuously (Fréchet) differentiable functions $X \to Y$, where $k \in \mathbb{N} \cup \{\infty\}$. By $\mathrm{C}^k(\overline{X};Y)$ we denote the subset of $\mathrm{C}^k(X;Y)$ of functions such that each of their derivatives up to order k can be continuously extended to the boundary of X.

In the case $Y = \mathbb{R}$ we simply write $\mathrm{C}(X)$, $\mathrm{C}^k(X)$, $\mathrm{C}^k(\overline{X})$ instead of $\mathrm{C}(X;\mathbb{R})$, $\mathrm{C}^k(X;\mathbb{R})$, $\mathrm{C}^k(\overline{X};\mathbb{R})$. By $\mathrm{C}_0^\infty(Z)$ we denote the set of all compactly supported functions in $\mathrm{C}^\infty(Z)$, where $Z \in \{X, \overline{X}\}$. The space of distributions, that is, the dual space of $\mathrm{C}_0^\infty(Z)$, is denoted by $\mathcal{D}'(Z)$.

Next we recall the notion of the Lebesgue integral. As customary, we always identify elements of these spaces, which are equivalence classes of functions, with one of their representatives and simply refer to them as functions as well.

Let X be a σ-finite measure space with measure μ. For $p \in [1, \infty)$ the symbol $\mathrm{L}^p(X)$ denotes the Lebesgue space of p-integrable functions, equipped with the norm

$$\|f\|_{p;X} := \left(\int_X |f|^p \, \mathrm{d}\mu \right)^{1/p},$$

and $\mathrm{L}^\infty(X)$ is the space of essentially bounded functions with norm

$$\|f\|_{\infty;X} := \mu\text{-}\operatorname*{ess\,sup}_{x \in X} |f(x)|.$$

Note that the notation does not indicate the underlying measure μ and does not distinguish between real-valued and complex-valued functions. In the following, both will always be clear from the context. If the underlying space X is also clear from the context, we simply write $\|\cdot\|_p$ instead of $\|\cdot\|_{p;X}$. Moreover, for $1 \le p < \infty$ the dual space of $\mathrm{L}^p(X)$ can be identified with $\mathrm{L}^{p'}(X)$, where p' is the Hölder conjugate of p defined $p' = p/(p-1)$ if $p \in (1, \infty)$, and $1' = \infty$ and $\infty' = 1$. We use this identification tacitly. Furthermore, $\mathrm{L}^p_{\mathrm{loc}}(X)$ denotes the set of all locally p-integrable functions on X. The weak Lebesgue space $\mathrm{L}^{p,\infty}(X)$ consists of all μ-measurable functions $f\colon X \to \mathbb{C}$ such that

$$\sup\{\alpha\,\mu(\{x \in X \mid |f(x)| > \alpha\})^{1/p} \mid \alpha > 0\} < \infty.$$

In the case $X = \mathbb{R}^n$, we further write $|\Omega|$ for the Lebesgue measure of a measurable set $\Omega \subset \mathbb{R}^n$. In the case $X = \mathbb{Z}$, which is always equipped with the counting measure, we set $\ell^p(\mathbb{Z}; X) := \mathrm{L}^p(\mathbb{Z}; X)$. If $\Omega \subset \mathbb{R}^n$ is a domain with boundary $\partial\Omega$, the symbol $\mathrm{d}S$ denotes its surface measure given by the restriction of the $(n-1)$-dimensional Hausdorff measure. If Ω has Lipschitz boundary, we denote by n its unit outer normal vector.

2.2 Harmonic Analysis on Groups

One advantage of the interpretation of modeling the time periodicity of functions by means of the torus group $\mathbb{T}$ is the availability of convolutions and a Fourier transform. In this section we prepare some related results from harmonic analysis on such locally compact abelian groups.

2.2.1 Convolutions

Let G be a locally compact abelian group equipped with the Haar measure μ_G. In our applications, G coincides with $\mathbb{R}^n$, $\mathbb{Z}$ or $\mathbb{T} = \mathbb{R}/\mathcal{T}\mathbb{Z}$ or a product of these spaces. Then $\mathbb{R}^n$ and $\mathbb{Z}$ are equipped with the standard Lebesgue measure $\mathrm{d}x$ and the counting measure, respectively, and $\mathbb{T}$ is equipped with the normalized Haar measure $\mathrm{d}t$ defined by

$$\forall\varphi \in \mathrm{C}(\mathbb{T}) : \quad \int_{\mathbb{T}} \varphi(t)\,\mathrm{d}t = \frac{1}{\mathcal{T}} \int_0^{\mathcal{T}} \varphi(t')\,\mathrm{d}t',$$

so that $\mathbb{T}$ is a finite measure space of measure 1.

The convolution $u * v$ of two functions $u, v\colon G \to \mathbb{C}$ is given by

$$u * v(x) := \int_G u(x-y)v(y)\,\mathrm{d}\mu_G(y),$$

provided that this is well defined. If different groups are involved, we also use the notation $u *_G v$ to specify the underlying group. Moreover, if $u, v\colon G \to \mathbb{C}^n$ are vector fields and $\Gamma\colon G \to \mathbb{C}^{n\times n}$ is a second-order tensor field, then the convolutions $u * v$ and $\Gamma * v$ are defined by $u * v := u_j * v_j$ and $(\Gamma * v)_j := \Gamma_{jk} * v_k$, where we employed the Einstein summation convention again.

2.2.2 The Fourier Transform

The symbol $\mathscr{S}(G)$ denotes the Schwartz–Bruhat space on G. This generalization of the classical Schwartz space was introduced by Bruhat [9]; for a precise definition we refer to [20]. For elementary groups of the form $G = \mathbb{Z}^\ell \times \mathbb{T}^m \times \mathbb{R}^n$ with $\ell, m, n \in \mathbb{N}_0$, which are those we encounter in the following, it is given by

$$\mathscr{S}(G) = \left\{ f \in C^\infty(G) \mid \forall \alpha, \beta \in \mathbb{N}_0^n, \gamma \in \mathbb{N}_0^m, \delta \in \mathbb{N}_0^\ell : \rho_{\alpha,\beta,\gamma,\delta}(f) < \infty \right\}$$

where

$$\rho_{\alpha,\beta,\gamma,\delta}(f) := \sup_{(k,t,x) \in \mathbb{Z}^\ell \times \mathbb{T}^m \times \mathbb{R}^n} |k^\delta x^\alpha D_x^\beta \partial_t^\gamma f(k,t,x)|.$$

Here $f \in C^\infty(G)$ means that $f(k, \cdot, \cdot) \in C^\infty(\mathbb{T}^m \times \mathbb{R}^n)$ for each $k \in \mathbb{Z}^\ell$. Equipped with the topology induced by these semi-norms, $\mathscr{S}(G)$ becomes a locally convex vector space. Its dual space $\mathscr{S}'(G)$ is called the space of tempered distributions on G. Observe that $\mathscr{S}(G)$ and $\mathscr{S}'(G)$ coincide with the classical Schwartz space and the space of tempered distributions in the Euclidean case $G = \mathbb{R}^n$.

As for classical tempered distributions in $\mathbb{R}^n$, one defines derivatives and multiplication via duality by

$$\langle D_x^\alpha D_t^\beta \Psi, \varphi \rangle := (-1)^{|\alpha|+|\beta|} \langle \Psi, D_x^\alpha D_t^\beta \varphi \rangle, \qquad \langle g\Psi, \varphi \rangle := \langle \Psi, g\varphi \rangle$$

for $\Psi \in \mathscr{S}'(G)$, $\varphi \in \mathscr{S}(G)$, $(\alpha, \beta) \in \mathbb{N}_0^n \times \mathbb{N}_0^m$ and a suitable smooth function $g : G \to \mathbb{C}$. Moreover, if there exists a function $f \in L^1_{\text{loc}}(G)$ such that

$$\langle \Psi, \varphi \rangle = \int_G f\varphi \, d\mu_G$$

for all $\varphi \in \mathscr{S}(G)$, then we call Ψ a regular distribution and identify it with f. In this way, the spaces $L^p(G)$, $p \in [1, \infty]$, are continuously embedded into $\mathscr{S}'(G)$.

The dual group $\widehat{G}$ of G consists of all continuous characters $\gamma : G \to \mathbb{T}$ on G and is equipped with the compact open topology. By the Pontryagin duality theorem, the dual group of $\widehat{G}$ can be identified with G. Then the Fourier transform $\mathscr{F}_G$ on G and its inverse $\mathscr{F}_G^{-1}$ are given by

$$\mathscr{F}_G : L^1(G) \to C(\widehat{G}), \qquad \mathscr{F}_G[f](\gamma) = \int_G f(x)\,\gamma(-x) \, d\mu_G(x),$$

$$\mathscr{F}_G^{-1} : L^1(\widehat{G}) \to C(G), \qquad \mathscr{F}_G^{-1}[f](x) = \int_{\widehat{G}} f(\gamma)\,\gamma(x) \, d\mu_{\widehat{G}}(\gamma).$$

The Fourier transform $\mathscr{F}_G$ is a continuous isomorphism $\mathscr{F}_G\colon \mathscr{S}(G) \to \mathscr{S}(\widehat{G})$, which extends to a continuous isomorphism $\mathscr{F}_G\colon \mathscr{S}'(G) \to \mathscr{S}'(\widehat{G})$ by duality, that is,

$$\langle \mathscr{F}_G[\Psi], \varphi \rangle := \langle \Psi, \mathscr{F}_{\widehat{G}}[\varphi] \rangle$$

for $\Psi \in \mathscr{S}'(G)$ and $\varphi \in \mathscr{S}(\widehat{G})$, which is justified by the Pontryagin duality theorem. If the Haar measures on G and $\widehat{G}$ are suitably normalized, then $\mathscr{F}_G$ is an isometric isomorphism $\mathscr{F}_G\colon L^2(G) \to L^2(\widehat{G})$ with inverse $\mathscr{F}_G^{-1}$.

One can identify the dual groups of $\mathbb{R}^n$ and $\mathbb{T}$ with $\widehat{\mathbb{R}^n} = \mathbb{R}^n$ and $\widehat{\mathbb{T}} = \mathbb{Z}$, respectively. More generally, the dual group of $G = \mathbb{Z}^\ell \times \mathbb{T}^m \times \mathbb{R}^n$ can be identified with $\widehat{G} = \mathbb{T}^\ell \times \mathbb{Z}^m \times \mathbb{R}^n$. This property allows to represent the corresponding Fourier transforms on $\mathscr{S}(G)$ as

$$\mathscr{F}_{\mathbb{T}}[f](k) = \int_{\mathbb{T}} f(t)\, e^{-i\frac{2\pi}{T}kt}\, dt, \qquad \mathscr{F}_{\mathbb{R}^n}[f](\xi) = \int_{\mathbb{R}^n} f(x)\, e^{-ix\cdot\xi}\, dx,$$

$$\mathscr{F}_{\mathbb{T}}^{-1}[f](t) = \sum_{k\in\mathbb{Z}} f(k)\, e^{i\frac{2\pi}{T}kt}, \qquad \mathscr{F}_{\mathbb{R}^n}^{-1}[f](x) = \int_{\mathbb{R}^n} f(\xi)\, e^{ix\cdot\xi}\, d\xi.$$

Due to $\mathscr{F}_{\mathbb{T}\times\mathbb{R}^n} = \mathscr{F}_{\mathbb{T}} \otimes \mathscr{F}_{\mathbb{R}^n}$ and $\mathscr{F}_{\mathbb{T}\times\mathbb{R}^n}^{-1} = \mathscr{F}_{\mathbb{T}}^{-1} \otimes \mathscr{F}_{\mathbb{R}^n}^{-1}$, we further obtain the identities

$$\mathscr{F}_{\mathbb{T}\times\mathbb{R}^n}[f](k,\xi) = \int_{\mathbb{T}}\int_{\mathbb{R}^n} f(t,x)\, e^{-ix\cdot\xi - i\frac{2\pi}{T}kt}\, dx dt,$$

$$\mathscr{F}_{\mathbb{T}\times\mathbb{R}^n}^{-1}[f](t,x) = \sum_{k\in\mathbb{Z}}\int_{\mathbb{R}^n} f(k,\xi)\, e^{ix\cdot\xi + i\frac{2\pi}{T}kt}\, d\xi.$$

Observe that the Lebesgue measure $d\xi$ on $\widehat{\mathbb{R}^n}$ has to be renormalized by a factor $(2\pi)^n$ in order to obtain an isometry $\mathscr{F}_{\mathbb{R}^n}\colon L^2(\mathbb{R}^n) \to L^2(\mathbb{R}^n)$. Moreover, the Fourier transform $\mathscr{F}_{\mathbb{T}}$ on $\mathbb{T}$ coincides with the classical Fourier expansion.

The behavior of derivatives under the Fourier transform is as in the classical Euclidean case. The identity

$$\mathscr{F}_{\mathbb{T}\times\mathbb{R}^n}\big[\partial_t^\ell D_x^\alpha \Psi\big] = \Big(i\frac{2\pi}{T}k\Big)^\ell (i\xi)^\alpha \mathscr{F}_{\mathbb{T}\times\mathbb{R}^n}[\Psi]$$

holds for all $\alpha \in \mathbb{N}_0^n$, $\ell \in \mathbb{N}_0$ and all tempered distributions $\Psi \in \mathscr{S}'(\mathbb{T}\times\mathbb{R}^n)$.

The delta distributions in $\mathbb{R}^n$ and $\mathbb{Z}$ are denoted by $\delta_{\mathbb{R}^n}$ and $\delta_{\mathbb{Z}}$, respectively. The support of a distribution Ψ is denoted by $\operatorname{supp}\Psi$.

2.2.3 Fourier Multipliers

A famous concept in the field of harmonic analysis is the theory of Fourier multipliers.

Definition 2.2.1. Let G be a locally compact abelian group and $p \in [1, \infty)$. We call $m \in \mathrm{L}^\infty(\widehat{G})$ an $\mathrm{L}^p(G)$ *multiplier* if the operator

$$\mathrm{op}_G[m] \colon \mathscr{S}(G) \to \mathscr{S}'(G), \quad f \mapsto \mathscr{F}_G^{-1}[m \mathscr{F}_G[f]],$$

satisfies

$$\forall f \in \mathscr{S}(G) \colon \quad \|\mathrm{op}_G[m](f)\|_{\mathrm{L}^p(G)} \leq C_1 \|f\|_{\mathrm{L}^p(G)}.$$

Then the operator $\mathrm{op}_G[m]$ has a continuous extension to an operator $\mathrm{L}^p(G) \to \mathrm{L}^p(G)$, and we simply write $\mathrm{op}_G[m] \in \mathcal{L}(\mathrm{L}^p(G))$. We also say that m is an L^p multiplier on G.

The question whether or not a function m is an L^p multiplier can be a hard one. While the set of $\mathrm{L}^2(G)$ multipliers coincides with $\mathrm{L}^\infty(\widehat{G})$ by Plancherel's Theorem, and the set of $\mathrm{L}^1(G)$ multipliers is given by the set of all regular finite Borel measures on G (see [77, Theorem 0.1.1] for example), an analogous characterization for general $p \in (1, \infty)$ is not known. However, there are several famous results, which nowadays belong to the standard repertoire in Fourier analysis and give sufficient conditions for a function m to be an L^p multiplier in the setting of the Euclidean space $G = \mathbb{R}^n$. One of these is the Marcinkiewicz Multiplier Theorem (see Theorem A.3.3).

However, on general groups such tools are not available directly, and one has to circumvent this issue. The method applied here is the idea of multiplier transference. In order to show that a function m is an L^p multiplier on one group, we transfer it to a different group, where better tools may be available. This idea goes back to DE LEEUW [16] and was further extended by EDWARDS and GAUDRY [18].

Theorem 2.2.2 (Transference Principle). *Let G and H be locally compact abelian groups, and let $\Phi \colon \widehat{H} \to \widehat{G}$ be a continuous homomorphism. Let $p \in (1, \infty)$ and*

$$m \in \mathrm{L}^\infty(\widehat{G}) \cap \mathrm{C}(\widehat{G})$$

be a continuous $\mathrm{L}^p(G)$ multiplier. Then $M := m \circ \Phi$ is an $\mathrm{L}^p(H)$ multiplier and

$$\|\mathrm{op}_H[M]\|_{\mathscr{L}(\mathrm{L}^p(H))} \leq \|\mathrm{op}_G[m]\|_{\mathscr{L}(\mathrm{L}^p(G))}. \tag{2.1}$$

Proof. See [18, Theorem B.2.1] or [20, Theorem 2.15]. □

Remark 2.2.3. In our applications, we usually meet two different cases. Either we have

$$G = \mathbb{R}, \qquad H = \mathbb{T}, \qquad \Phi\colon \mathbb{Z} \to \mathbb{R}, \ k \mapsto k,$$

or

$$G = \mathbb{R} \times \mathbb{R}^n, \qquad H = \mathbb{T} \times \mathbb{R}^n, \qquad \Phi\colon \mathbb{Z} \times \mathbb{R}^n \to \mathbb{R} \times \mathbb{R}^n, \ (k, \xi) \mapsto (k, \xi).$$

In both cases, Φ is the trivial embedding, so that $M = m|_H$, that is, M is the restriction of m to H, which is a null set in G. This can be seen as the justification for the continuity of m required in Theorem 2.2.2.

As an example, we show the continuity of the Riesz transform in $L^p(\mathbb{T})$.

Proposition 2.2.4. *The Riesz transform $\mathfrak{R}_\mathbb{T}$ in the torus $\mathbb{T}$ given by*

$$\mathfrak{R}_\mathbb{T}\colon \mathscr{S}(\mathbb{T}) \to \mathscr{S}'(\mathbb{T}), \qquad \mathfrak{R}_\mathbb{T}(f) \coloneqq \mathscr{F}_\mathbb{T}^{-1}\big[-i\,\mathrm{sgn}(k)\mathscr{F}_\mathbb{T}[f]\big] \qquad (2.2)$$

can be extended to a continuous linear operator $L^p(\mathbb{T}) \to L^p(\mathbb{T})$ for any $p \in (1, \infty)$.

Proof. Let $M(k) \coloneqq -i\,\mathrm{sgn}(k)$. Then the statement follows if M is an $L^p(\mathbb{T})$ multiplier for any $p \in (1, \infty)$. Let $\chi \in C_0^\infty(\mathbb{R})$ be a cut-off function with $\chi(\eta) = 1$ for $|\eta| \leq \frac{1}{2}$ and $\chi(\eta) = 0$ for $|\eta| \geq 1$. Set $m(\eta) \coloneqq -i(1-\chi(\eta))\,\mathrm{sgn}(\eta)$. Then m is smooth and $m'(\eta) = i\chi'(\eta)\,\mathrm{sgn}(\eta)$. Therefore, m and m' are uniformly bounded, and the Marcinkiewicz Multiplier Theorem (Theorem A.3.3) shows that m is an $L^p(\mathbb{R})$ multiplier. Since $M = m|_\mathbb{Z}$, we conclude the proof by the Transference Principle (Theorem 2.2.2). □

2.3 Sobolev Spaces

Here we introduce the notation for classical and homogeneous Sobolev spaces. At first, we consider functions that only depend on spatial variables, and we recall the famous Gagliardo–Nirenberg inequality. We introduce spaces of functions that also depend on time and collect embedding results for time-periodic functions.

2.3.1 Classical and Homogeneous Sobolev Spaces

Let Ω be an open subset of $\mathbb{R}^n$, $n \in \mathbb{N}$, and let $k \in \mathbb{N}_0$ and $p \in [1, \infty]$. By $W^{k,p}(\Omega)$ we denote the inhomogeneous Sobolev space that contains all functions $u \in L^p(\Omega)$ with weak derivatives in $D^\alpha u \in L^p(\Omega)$ for all $|\alpha| \leq k$, which is equipped with the norm

$$\|u\|_{k,p;\Omega} := \sum_{|\alpha| \leq k} \|D^\alpha u\|_{p;\Omega}.$$

For $A \in \{\Omega, \overline{\Omega}\}$, the space $W^{k,p}_{\mathrm{loc}}(A)$ contains all functions u such that $D^\alpha u \in L^p_{\mathrm{loc}}(A)$ for all $|\alpha| \leq k$. Observe that $W^{k,p}_{\mathrm{loc}}(\Omega) \neq W^{k,p}_{\mathrm{loc}}(\overline{\Omega})$ in general. Moreover, the homogeneous Sobolev space $D^{k,p}(\Omega)$ is defined by

$$D^{k,p}(\Omega) := \left\{ u \in L^1_{\mathrm{loc}}(\Omega) \mid D^\alpha u \in L^p(\Omega) \text{ for all } |\alpha| = k \right\},$$

which we equip with the semi-norm

$$|u|_{k,p;\Omega} := \sum_{|\alpha| = k} \|D^\alpha u\|_{p;\Omega}.$$

When the underlying domain Ω is clear, we simply write $\|\cdot\|_{k,p}$ and $|\cdot|_{k,p}$ instead of $\|\cdot\|_{k,p;\Omega}$ and $|\cdot|_{k,p;\Omega}$.

In contrast to $W^{k,p}(\Omega)$, the space $D^{k,p}(\Omega)$ is not a Banach space for $k \geq 1$ since $|\cdot|_{k,p}$ is not definite in this case. However, both $\|\cdot\|_{k,p}$ and $|\cdot|_{k,p}$ are norms on $C_0^\infty(\Omega)$, so that one can define the respective Cantor completions

$$W^{k,p}_0(\Omega) := \overline{C_0^\infty(\Omega)}^{\|\cdot\|_{k,p}}, \qquad D^{k,p}_0(\Omega) := \overline{C_0^\infty(\Omega)}^{|\cdot|_{k,p}}.$$

For $p \in (1, \infty)$ their dual spaces are denoted by

$$W^{-k,p}_0(\Omega) := \left(W^{k,p'}_0(\Omega) \right)', \qquad D^{-k,p}_0(\Omega) := \left(D^{k,p'}_0(\Omega) \right)',$$

where $p' = p/(p-1)$ is the Hölder conjugate of p. Their norms are denoted by $\|\cdot\|_{-k,p} = \|\cdot\|_{-k,p;\Omega}$ and $|\cdot|_{-k,p} = |\cdot|_{-k,p;\Omega}$.

At this point, we prepare the following density result.

Proposition 2.3.1. *Let $\Omega \subset \mathbb{R}^n$ be an arbitrary domain and $q, r \in (1, \infty)$. Then $C_0^\infty(\Omega)$ is a dense subset of $L^q(\Omega) \cap D_0^{-1,r}(\Omega)$.*

Proof. The space $L^q(\Omega) \cap D_0^{-1,r}(\Omega)$ can be identified with the dual space of $L^{q'}(\Omega) + D_0^{1,r'}(\Omega)$, where $s' = s/(s-1)$ for $s \in \{q, r\}$. Identifying elements of

$C_0^\infty(\Omega)$ as regular distributions in $\mathcal{D}'(\Omega)$, we consider $g \in L^{q'}(\Omega) + D_0^{1,r'}(\Omega)$ belonging to the kernel of each functional in $C_0^\infty(\Omega)$, that is, such that

$$\int_\Omega \varphi g \, dx = 0$$

for all $\varphi \in C_0^\infty(\Omega)$. This implies $g = 0$. Consequently, by a standard duality argument, $C_0^\infty(\Omega)$ is dense in $L^q(\Omega) \cap D_0^{-1,r}(\Omega)$. $\qquad\square$

The following theorem gives an estimate of the boundary trace of a function in $W^{1,2}(\Omega)$. We only consider a very particular case here, which is sufficient for our applications.

Theorem 2.3.2. *Let $\Omega \subset \mathbb{R}^n$ be a domain with Lipschitz boundary. For every $\varepsilon > 0$ there exists a constant $C_2 = C_2(n, \Omega, \varepsilon) > 0$ such that the estimate*

$$\|u\|_{2;\partial\Omega} \le \varepsilon \|\nabla u\|_{2;\Omega} + C_2 \|u\|_{2;\Omega}.$$

holds for all $u \in W^{1,2}(\Omega)$.

Proof. By a classical trace inequality (see [42, Theorem II.4.1] for example) we have

$$\|u\|_{2;\partial\Omega} \le c_0 \|u\|_{2;\Omega}^{\frac{1}{2}} \|u\|_{1,2;\Omega}^{\frac{1}{2}} \le c_1 \|u\|_{2;\Omega}^{\frac{1}{2}} \left(\|u\|_{2;\Omega} + \|\nabla u\|_{2;\Omega}\right)^{\frac{1}{2}}$$

$$\le c_2\left(\|u\|_{2;\Omega} + \|u\|_{2;\Omega}^{\frac{1}{2}} \|\nabla u\|_{2;\Omega}^{\frac{1}{2}}\right).$$

Now the assertion follows by an application of Young's inequality. $\qquad\square$

The following proposition is a generalization of Poincaré's inequality to second-order derivatives.

Proposition 2.3.3 (Second-Order Poincaré Inequality). *Let $\Omega \subset \mathbb{R}^n$, $n \ge 2$, be a bounded Lipschitz domain, and let Γ be an $(n-1)$-dimensional connected component of $\partial\Omega$. For $1 < q < \infty$ there exists a constant $C_3 = C_3(n, \Omega, \Gamma, q) > 0$ such that*

$$\|u\|_q + \|\nabla u\|_q \le C_3 \|\nabla^2 u\|_q \tag{2.3}$$

for all $u \in W^{2,q}(\Omega)$ with $u = 0$ on Γ.

Proof. We prove the statement by a contradiction argument and assume that it does not hold. Then there exists a sequence $(u_n) \subset W^{2,q}(\Omega)$ with $u_n = 0$ on Γ such that

$$\|u_n\|_q + \|\nabla u_n\|_q = 1, \qquad \|\nabla^2 u_n\|_q \leq \frac{1}{n}.$$

Then the sequence (u_n) is uniformly bounded in $W^{2,q}(\Omega)$ and thus contains a weakly convergent subsequence (which we identify with (u_n)) with limit $u \in W^{2,q}(\Omega)$ satisfying $u = 0$ on Γ. By the compact embedding $W^{2,q}(\Omega) \hookrightarrow W^{1,q}(\Omega)$, the function u is the strong limit of (u_n) in $W^{1,q}(\Omega)$. In particular, we conclude

$$\|u\|_q + \|\nabla u\|_q = 1. \tag{2.4}$$

Because $(\nabla^2 u_n)$ converges to 0 by assumption, (u_n) converges strongly to u in $W^{2,q}(\Omega)$ and $\nabla^2 u = 0$. This implies $u(x) = a \cdot x + b$ for $a \in \mathbb{R}^n$, $b \in \mathbb{R}$. We set

$$S_{a,b} := \{x \in \mathbb{R}^n \mid a \cdot x + b = 0\}.$$

Since $u = 0$ on Γ, the hypersurface Γ belongs to the affine linear space $S_{a,b}$. By our assumption, this is only possible if $S_{a,b} = \mathbb{R}^n$, which is equivalent to $a = 0$ and $b = 0$. This implies $u = 0$, which contradicts (2.4) and completes the proof. □

Remark 2.3.4. Observe that the assumptions on Γ in Proposition 2.3.3 can be weakened. As we see from the proof, it suffices to assume that Γ is a subset of $\partial\Omega$ such that the vanishing-trace condition $u|_\Gamma = 0$ makes sense and Γ is not contained in a proper affine subspace of $\mathbb{R}^n$.

Moreover, Proposition 2.3.3 can directly be generalized to higher-order derivatives by imposing additional geometric requirements on Γ. For example, an estimate of the form $\|u\|_{2,q} \leq C_4 \|\nabla^3 u\|_q$ holds under the additional assumption that Γ is not contained in a quadric hypersurface.

2.3.2 Interpolation Inequalities

Next we recall the famous Gagliardo–Nirenberg inequality in bounded and exterior domains and conclude a simple corollary.

Theorem 2.3.5 (Gagliardo–Nirenberg inequality). *Let $\Omega \subset \mathbb{R}^n$, $n \geq 2$, be a bounded domain with Lipschitz boundary, and let $p \in (1, \infty]$, $q \in (1, \infty)$,*

$m \in \mathbb{N}$ and $u \in D^{m,p}(\Omega) \cap L^q(\Omega)$. Moreover, let $k \in \mathbb{N}_0$, $0 \le k < m$. Then $u \in D^{k,r}(\Omega)$ and

$$|u|_{k,r} \le C_5\big(|u|_{m,p}^{\theta}\|u\|_q^{1-\theta} + \|u\|_q\big),$$

where

$$\frac{1}{r} = \frac{k}{n} + \theta\left(\frac{1}{p} - \frac{m}{n}\right) + (1-\theta)\frac{1}{q}$$

with $\theta \in [k/m, 1)$, and $C_5 = C_5(n, \Omega, p, q, m, k, \theta) > 0$. If $m - k - n/p \notin \mathbb{N}_0$, then $\theta = 1$ is also admissible.

Proof. See [37, 84]. $\qquad\square$

The following generalized Ehrling's inequality is a rather direct consequence.

Corollary 2.3.6. *Let $\Omega \subset \mathbb{R}^n$, $n \ge 2$, be a bounded domain with Lipschitz boundary, $N \in \mathbb{N}$ and p, $q_k \in (1, \infty)$, $k = 0, \ldots, N-1$ and $\varepsilon > 0$. If $u \in W^{N,p}(\Omega)$ and $u \in W^{k,q_k}(\Omega)$ for $k = 1, \ldots, N-1$, then*

$$\|u\|_p \le C_6 \sum_{k=0}^{N-1} |u|_{k,q_k} + \varepsilon|u|_{m,p},$$

where $C_6 = C_6(\varepsilon, n, \Omega, N, p, q_0, \ldots, q_{N-1}) > 0$.

Proof. First we show the estimate for $N = 1$. Assume $u \in W^{1,p}(\Omega) \cap L^q(\Omega)$ for $p, q \in (1, \infty)$. If $q \ge p$, since Ω is a bounded domain, we have

$$\|u\|_p \le c_0\|u\|_q \le c_0\|u\|_q + \varepsilon\|u\|_{1,p}$$

for any $\varepsilon > 0$. If $q < p$, we employ the Gagliardo–Nirenberg inequality (Theorem 2.3.5) with $\theta = n(p-q)/(np - nq + pq)$ to obtain

$$\|u\|_p \le c_1\big(|u|_{1,p}^{\theta}\|u\|_q^{1-\theta} + \|u\|_q\big) \le c_2(1+\varepsilon)\|u\|_q + \varepsilon|u|_{1,p}$$

by Young's inequality. This completes the proof in the case $N = 1$. The general case now follows iteratively. $\qquad\square$

The following result generalizes Theorem 2.3.5 to the case of an exterior domain.

Theorem 2.3.7 (Gagliardo–Nirenberg inequality). *Let $\Omega \subset \mathbb{R}^n$, $n \ge 2$, be an exterior domain with Lipschitz boundary or $\Omega = \mathbb{R}^n$. Let $p \in (1, \infty]$,*

$q \in (1, \infty)$, $m \in \mathbb{N}$ and $u \in D^{m,p}(\Omega) \cap L^q(\Omega)$. *Moreover, let* $k \in \mathbb{N}_0$, $0 \le k < m$. *Then* $u \in D^{k,r}(\Omega)$ *and*

$$|u|_{k,r} \le C_7 |u|_{m,p}^{\theta} \|u\|_q^{1-\theta},$$

where

$$\frac{1}{r} = \frac{k}{n} + \theta \left(\frac{1}{p} - \frac{m}{n} \right) + (1 - \theta) \frac{1}{q}$$

with $\theta \in [k/m, 1)$, *and* $C_7 = C_7(n, \Omega, p, q, m, k, \theta) > 0$. *If* $m - k - n/p \notin \mathbb{N}_0$, *then* $\theta = 1$ *is also admissible.*

Proof. See [15]. □

2.3.3 Functions of Space and Time

Let $\Omega \subset \mathbb{R}^n$ and $f \in L^1_{loc}(\mathbb{T} \times \Omega)$, and let $X(\Omega)$ be a semi-normed function space. For $p \in [1, \infty]$, we write $f \in L^p(\mathbb{T}; X(\Omega))$ if the function

$$t \mapsto \|f(t, \cdot)\|_{X(\Omega)}$$

belongs to $L^p(\mathbb{T})$, and for $p \in [1, \infty)$ we set

$$\|f\|_{L^p(\mathbb{T};X(\Omega))} := \left(\int_{\mathbb{T}} \|f(t, \cdot)\|_{X(\Omega)}^p \, dt \right)^{1/p},$$

$$\|f\|_{L^\infty(\mathbb{T};X(\Omega))} := \operatorname{ess\,sup}_{t \in \mathbb{T}} \|f(t, \cdot)\|_{X(\Omega)}.$$

We further introduce the *steady-state projection*

$$\mathcal{P}f(x) := \int_{\mathbb{T}} f(t, x) \, dt, \qquad \mathcal{P}_\perp f := f - \mathcal{P}f.$$

Observe that $\mathcal{P}f$ is time-independent. Therefore, we call $\mathcal{P}f$ *steady-state* part and $\mathcal{P}_\perp f$ *purely periodic* part of f.

The operators $\mathcal{P}$ and $\mathcal{P}_\perp$ are continuous projections $L^p(\mathbb{T}; X(\Omega)) \to L^p(\mathbb{T}; X(\Omega))$, which leads to the direct decomposition

$$L^p(\mathbb{T}; X(\Omega)) = X(\Omega) \oplus L^p_\perp(\mathbb{T}; X(\Omega)),$$

where we set $L^p_\perp(\mathbb{T}; X(\Omega)) := \mathcal{P}_\perp L^p(\mathbb{T}; X(\Omega))$. If $X(\Omega) = L^p(\Omega)$, we define $L^p_\perp(\mathbb{T} \times \Omega) := \mathcal{P}_\perp L^p(\mathbb{T} \times \Omega)$.

We further introduce the space

$$W^{1,2,p}(\mathbb{T} \times \Omega) := \{u \in L^p(\mathbb{T}; W^{2,p}(\Omega)) \mid \partial_t u \in L^p(\mathbb{T} \times \Omega)\},$$

which is equipped with the norm

$$\|u\|_{1,2,p} := \sum_{|\alpha| \le 2} \|D_x^\alpha u\|_p + \|\partial_t u\|_p,$$

and its purely periodic subspace $W_\perp^{1,2,q}(\mathbb{T} \times \Omega) := \mathcal{P}_\perp W^{1,2,q}(\mathbb{T} \times \Omega)$.

Moreover, for $G = \mathbb{T} \times \mathbb{R}^n$ the operators $\mathcal{P}$ and $\mathcal{P}_\perp$ are continuous mappings $\mathscr{S}(G) \to \mathscr{S}(G)$. They can thus be transferred to mappings on $\mathscr{S}'(G)$ by duality via

$$\langle \mathcal{P}\Psi, \varphi \rangle := \langle \Psi, \mathcal{P}\varphi \rangle$$

for $\Psi \in \mathscr{S}'(G)$ and $\varphi \in \mathscr{S}(G)$. As before, we obtain direct decompositions of $\mathscr{S}(G)$ and $\mathscr{S}'(G)$.

Observe that $\mathcal{P}f = \mathscr{F}_{\mathbb{T}}[f](0) = \mathscr{F}_{\mathbb{T}}^{-1}[\delta_{\mathbb{Z}}\mathscr{F}_{\mathbb{T}}[f]]$ since $\mathscr{F}_{\mathbb{T}}^{-1}[\delta_{\mathbb{Z}}] = 1$. This further yields the identity

$$\mathcal{P}_\perp f = \mathscr{F}_{\mathbb{T}}^{-1}[(1 - \delta_{\mathbb{Z}})\mathscr{F}_{\mathbb{T}}[f]].$$

2.3.4 Embedding Theorems

Here we consider embedding properties of time-periodic functions, more specifically, of functions in $W^{1,2,q}(\mathbb{T} \times \Omega)$. The following theorem is due to GALDI and KYED [50].

Theorem 2.3.8. *Let $\Omega \subset \mathbb{R}^n$, $n \ge 2$, be the whole space $\mathbb{R}^n$ or a bounded or exterior domain in $\mathbb{R}^n$ with Lipschitz boundary, and let $q \in (1, \infty)$. Assume that $\alpha \in [0, 2]$ and $p_0, r_0 \in [q, \infty]$ satisfy*

$$\begin{cases} r_0 \le \dfrac{2q}{2 - \alpha q} & \text{if } \alpha q < 2, \\ r_0 < \infty & \text{if } \alpha q = 2, \\ r_0 \le \infty & \text{if } \alpha q > 2, \end{cases} \qquad \begin{cases} p_0 \le \dfrac{nq}{n - (2 - \alpha)q} & \text{if } (2 - \alpha)q < n, \\ p_0 < \infty & \text{if } (2 - \alpha)q = n, \\ p_0 \le \infty & \text{if } (2 - \alpha)q > n, \end{cases}$$

and that $\beta \in [0, 1]$ and $p_1, r_1 \in [q, \infty]$ satisfy

$$\begin{cases} r_1 \le \dfrac{2q}{2 - \beta q} & \text{if } \beta q < 2, \\ r_1 < \infty & \text{if } \beta q = 2, \\ r_1 \le \infty & \text{if } \beta q > 2, \end{cases} \qquad \begin{cases} p_1 \le \dfrac{nq}{n - (1 - \beta)q} & \text{if } (1 - \beta)q < n, \\ p_1 < \infty & \text{if } (1 - \beta)q = n, \\ p_1 \le \infty & \text{if } (1 - \beta)q > n. \end{cases}$$

Then there is a constant $C_8 = C_8(n, \Omega, \mathcal{T}, q, r_0, p_0, r_1, p_1) > 0$ such that the inequality

$$\|u\|_{L^{r_0}(\mathbb{T};L^{p_0}(\Omega))} + \|\nabla u\|_{L^{r_1}(\mathbb{T};L^{p_1}(\Omega))} \leq C_8 \|u\|_{1,2,q} \tag{2.5}$$

holds for all $u \in W^{1,2,q}(\mathbb{T} \times \Omega)$.

Proof. This result was proved in [50, Theorem 4.1] for exterior domains of class C^1. However, the proof is first established on the whole space and then transferred to an exterior domain by means of classical Sobolev extension operators. Since these operators exist also for bounded and exterior domains with Lipschitz boundary, the generalization to these domains is straightforward. □

We also need the following refinement of Theorem 2.3.8, which takes into account a weight in front of the time derivative. Clearly, this does not affect estimates of the steady-state part of a function $u \in W^{1,2,q}(\mathbb{T} \times \Omega)$, which is why we only consider the case of purely periodic functions.

Theorem 2.3.9. *Let $n \geq 2$, $\omega > 0$ and $q \in (1, \infty)$. For $\alpha \in [0, 2]$ with $\alpha q < 2$ and $(2 - \alpha)q < n$ let*

$$r_0 := \frac{2q}{2 - \alpha q}, \qquad p_0 := \frac{nq}{n - (2 - \alpha)q},$$

and for $\beta \in [0, 1]$ with $\beta q < 2$ and $(1 - \beta)q < n$ let

$$r_1 := \frac{2q}{2 - \beta q}, \qquad p_1 := \frac{nq}{n - (1 - \beta)q}.$$

Then there is a constant $C_9 = C_9(n, \mathcal{T}, q, \alpha, \beta) > 0$ such that the inequality

$$\omega^{\alpha/2}\|u\|_{L^{r_0}(\mathbb{T};L^{p_0}(\mathbb{R}^n))} + \omega^{\beta/2}\|\nabla u\|_{L^{r_1}(\mathbb{T};L^{p_1}(\mathbb{R}^n))}$$
$$\leq C_9\big(\omega\|\partial_t u\|_q + \|\nabla^2 u\|_q\big) \tag{2.6}$$

holds for all $u \in \mathcal{P}_\perp W^{1,2,q}(\mathbb{T} \times \mathbb{R}^n)$.

Proof. Since $\mathscr{S}(G)$ is dense in $W^{1,2,q}(G)$, it suffices to consider $u \in \mathscr{S}(G)$ with $\mathcal{P}u = 0$. In particular, we have $\mathscr{F}_G[u] = (1 - \delta_\mathbb{Z})\mathscr{F}_G[u]$. By means of the Fourier transform $\mathscr{F}_G$, we thus derive the identity

$$u = \mathscr{F}_G^{-1}\left[\frac{1 - \delta_\mathbb{Z}(k)}{|\xi|^2 + i\omega\frac{2\pi}{\mathcal{T}}k}\mathscr{F}_G[\omega\partial_t u - \Delta u]\right]$$
$$= \mathscr{F}_{\mathbb{R}^n}^{-1}\big[|\xi|^{\alpha - 2}\big] *_{\mathbb{R}^n} [\varphi_{\alpha/2} *_\mathbb{T} F], \tag{2.7}$$

where

$$F \coloneqq \mathscr{F}_G^{-1}\Big[M_\omega(k,\xi) \mathscr{F}_G\big[\omega \partial_t u - \Delta u \big] \Big],$$

$$M_\omega(k,\xi) \coloneqq \frac{|k|^{\alpha/2}|\xi|^{2-\alpha}(1 - \delta_{\mathbb{Z}}(k))}{|\xi|^2 + i\omega\frac{2\pi}{T}k},$$

$$\varphi_{\alpha/2} \coloneqq \mathscr{F}_{\mathbb{T}}^{-1}\Big[k \mapsto (1 - \delta_{\mathbb{Z}}(k))|k|^{-\alpha/2} \Big].$$

Then we have

$$M_\omega = \Big(\frac{2\pi}{T}\omega\Big)^{-\alpha/2} \widetilde{m}_{\kappa,\lambda}\big|_{\mathbb{Z}\times\mathbb{R}^n}$$

for $\widetilde{m}_{\kappa,\lambda}$ as in (A.87) and $\kappa = \omega\frac{2\pi}{T}$, $\lambda = 0$ and $\theta = \alpha/2$. By Lemma A.3.10, the function $\widetilde{m}_{\kappa,\lambda}$ is a continuous L^q multiplier on $\mathbb{R} \times \mathbb{R}^n$. In view of Remark 2.2.3, the Transference Principle (Theorem 2.2.2) thus shows that M_ω is an L^q multiplier on G and satisfies

$$\|\mathrm{op}_G[M_\omega]\|_{\mathcal{L}(L^q(G))} \le \Big(\frac{2\pi}{T}\omega\Big)^{-\alpha/2} \|\mathrm{op}_{\mathbb{R}\times\mathbb{R}^n}[\widetilde{m}_{\kappa,\lambda}]\|_{\mathcal{L}(L^q(\mathbb{R}\times\mathbb{R}^n))} \le c_0\omega^{-\alpha/2},$$

where the constant c_0 is independent of ω due to (A.90) and $\lambda = 0$. Consequently, we have

$$\omega^{\alpha/2}\|F\|_q \le c_0\|\omega\partial_t u - \Delta u\|_q \le c_0\big(\omega\|\partial_t u\|_q + \|\nabla^2 u\|_q\big).$$

Moreover, Lemma A.3.1 yields $\varphi_{\alpha/2} \in L^{\frac{1}{1-\alpha/2},\infty}(\mathbb{T})$. Furthermore, it is well known that the mapping $\varphi \mapsto \mathscr{F}_{\mathbb{R}^n}^{-1}\big[|\xi|^{\alpha-2}\big] *_{\mathbb{R}^n} \varphi$ extends to a bounded operator $L^q(\mathbb{R}^n) \to L^{p_0}(\mathbb{R}^n)$; see [57, Theorem 6.1.3] for example. Recalling (2.7) and employing the inequalities by Minkowski and Young, as $r_0 > q$ we have

$$\omega^{\alpha/2}\|u\|_{L^{r_0}(\mathbb{T};L^{p_0}(\mathbb{R}^n))} = \omega^{\alpha/2}\bigg(\int_{\mathbb{T}} \Big\| \mathscr{F}_{\mathbb{R}^n}^{-1}\big[|\xi|^{\alpha-2}\big] *_{\mathbb{R}^n} \varphi_{\alpha/2} *_{\mathbb{T}} F(t,\cdot) \Big\|_{p_0}^{r_0} \, dt\bigg)^{\frac{1}{r_0}}$$

$$\le c_1\omega^{\alpha/2}\bigg(\int_{\mathbb{T}} \|\varphi_{\alpha/2} *_{\mathbb{T}} F(t,\cdot)\|_q^{r_0} \, dt\bigg)^{\frac{1}{r_0}} \le c_2\omega^{\alpha/2}\bigg(\int_{\mathbb{R}^n} \|\varphi_{\alpha/2} *_{\mathbb{T}} F(\cdot,x)\|_{r_0}^q \, dx\bigg)^{\frac{1}{q}}$$

$$\le c_3\omega^{\alpha/2}\|F\|_q \le c_4\big(\omega\|\partial_t u\|_q + \|\nabla^2 u\|_q\big).$$

This is the asserted inequality for u. The estimate of ∇u follows in the same way. $\qquad\square$

Remark 2.3.10. Note that the term on the right-hand side of (2.6) defines a norm equivalent to $\|\cdot\|_{1,2,q}$ on $W_{\perp}^{1,2,q}(\mathbb{T}\times\mathbb{R}^n)$. More precisely, since $\mathcal{P}u = 0$ for $u \in W_{\perp}^{1,2,q}(\mathbb{T}\times\mathbb{R}^n)$, Poincaré's inequality yields

$$\omega\|u\|_q \le C_{10}\omega\|\partial_t u\|_q.$$

By the Gagliardo–Nirenberg inequality (Theorem 2.3.7), this implies

$$C_{11}^{-1}\|u\|_{1,2,q} \le \omega\|\partial_t u\|_q + \|\nabla^2 u\|_q \le C_{11}\|u\|_{1,2,q}$$

for a constant $C_{11} > 0$ depending on ω.

Remark 2.3.11. Theorem 2.3.9 can be generalized to the setting of an exterior domain $\Omega \subset \mathbb{R}^n$ by means of Sobolev extensions. However, in order to maintain the homogeneous estimate (2.6), one has to construct a specific extension operator that respects the homogeneous second-order Sobolev norm. To this end, one can make use of results from [10].

2.4 Mathematical Fluid Dynamics

In this section we present some results from mathematical fluid dynamics that nowadays belong to the standard theory in this field.

2.4.1 The Helmholtz–Weyl Decompositon

The Helmholtz–Weyl decomposition is used to split a vector field into a divergence-free (also called solenoidal) part and a gradient field. For its definition, let $\Omega \subset \mathbb{R}^n$, $n \ge 2$, be a domain. Then $C_{0,\sigma}^{\infty}(\Omega)$ denotes the space of all divergence-free smooth vector fields with compact support, that is,

$$C_{0,\sigma}^{\infty}(\Omega) := \left\{\varphi \in C_0^{\infty}(\Omega)^n \mid \operatorname{div}\varphi = 0\right\}.$$

For $q \in (1,\infty)$, the space $L_{\sigma}^q(\Omega)$ of all solenoidal vector fields in $L^q(\Omega)$ and the space $\mathscr{G}^q(\Omega)$ of all gradient fields in $L^q(\Omega)$ are defined by

$$L_{\sigma}^q(\Omega) := \overline{C_{0,\sigma}^{\infty}(\Omega)}^{\|\cdot\|_q}, \qquad \mathscr{G}^q(\Omega) := \left\{\nabla\mathfrak{p} \mid \mathfrak{p} \in D^{1,q}(\Omega)\right\}.$$

Then the following theorem collects famous results.

Theorem 2.4.1. *Let $\Omega = \mathbb{R}^n$ or let $\Omega \subset \mathbb{R}^n$ be a bounded or exterior domain of class C^2. Let $q \in (1,\infty)$. For each $f \in L^q(\Omega)$ there are unique* *elements $u \in L_\sigma^q(\Omega)$ and $g = \nabla\mathfrak{p} \in \mathscr{G}^q(\Omega)$ such that $f = u + g = u + \nabla\mathfrak{p}$. In other words, the unique decomposition*

$$L^q(\Omega) = L_\sigma^q(\Omega) \oplus \mathscr{G}^q(\Omega),$$

called Helmholtz–Weyl *decomposition, holds. Moreover, there exists an associated projection operator $\mathcal{P}_H$, called* Helmholtz projector, *such that $\mathcal{P}_H f = u$. Then $\mathcal{P}_H : L^q(\Omega) \to L^q(\Omega)$ is a continuous linear operator that satisfies $\mathcal{P}_H^2 = \mathcal{P}_H$, and $\mathcal{P}_H$ is an orthogonal projection if $q = 2$.*

Proof. A proof can be found in [42, Theorem III.1.2] for example. □

Note that in the case $q = 2$ the Helmholtz–Weyl decomposition is valid for any domain Ω; see [42, Theorem III.1.1] for example. In contrast, for any $q \neq 2$ there exist domains, where this decomposition does not hold; see further references in [42, Section III.1].

We further introduce the homogeneous space

$$D_{0,\sigma}^{1,q}(\Omega) := \overline{C_{0,\sigma}^\infty(\Omega)}^{|\cdot|_{1,q}}.$$

2.4.2 The Divergence Problem

Here we collect famous results about the divergence problem, that is, given a function $g: \Omega \to \mathbb{R}$ on a domain $\Omega \subset \mathbb{R}^n$, to find a function $v: \Omega \to \mathbb{R}^n$ such that

$$\begin{cases} \operatorname{div} v = g & \text{in } \Omega, \\ \quad v = 0 & \text{on } \partial\Omega, \end{cases} \tag{2.8}$$

and satisfying suitable *a priori* estimates.

We shall often encounter this problem in the following setting. When deriving properties for a function $u: \Omega \to \mathbb{R}^n$ defined on an exterior domain Ω, we frequently multiply u with a cut-off function $\chi \in C^\infty(\mathbb{R}^n)$ with specific support and obtain a function $w := \chi u$. When u satisfies $\operatorname{div} u = 0$, this property is lost during this procedure in general, and $\operatorname{div} w$ does not vanish everywhere since $\operatorname{div} w = u \cdot \nabla\chi$. To obtain a divergence-free function again, the idea is to subtract a function v satisfying (2.8) with $g = u \cdot \nabla\chi$.

In a bounded domain, this problem is resolved by the Bogovskiĭ operator $\mathfrak{B}$. Before its introduction, note that if Ω is a bounded domain, the assumption

$$\int_\Omega g \, dx = 0 \tag{2.9}$$

is necessary for the existence of a solution v to (2.8) due to the divergence theorem.

Theorem 2.4.2 (Bogovskiĭ Operator). *Let $D \subset \mathbb{R}^n$, $n \geq 2$ be a bounded domain with Lipschitz boundary, There is a linear operator $\mathfrak{B} \colon C_0^\infty(D) \to C_0^\infty(D)^n$, called* Bogovskiĭ operator, *with the property*

$$\int_D \varphi \, dx = 0 \quad \Longrightarrow \quad \operatorname{div} \mathfrak{B}\varphi = \varphi.$$

For $q \in (1, \infty)$ and $m \in \mathbb{N}_0$, this operator has a continuous extension to a linear operator $\mathfrak{B} \colon W_0^{m,q}(D) \to W_0^{m+1,q}(D)^n$. Moreover, there exists a constant $C_{12} = C_{12}(n, D, q) > 0$ such that

$$\|\mathfrak{B}g\|_q \leq C_{12} |g|_{-1,q;D}^* \tag{2.10}$$

for all $g \in L^q(D)$, where $|\cdot|_{-1,q;D}^*$ is defined by

$$|g|_{-1,q;D}^* := \sup\left\{ \left| \int_D g\psi \, dx \right| \;\middle|\; \psi \in C_0^\infty(\mathbb{R}^n), \; \|\nabla\psi\|_{q';D} = 1 \right\}. \tag{2.11}$$

In particular, if g satisfies (2.9), then $v = \mathfrak{B}g$ is a solution to (2.8).

Proof. See [42, Theorem III.3.3 and Theorem III.3.5] for example. $\qquad\square$

Let us have a closer look at estimate (2.10). The term $|g|_{-1,q;D}^*$ on the right-hand side seems unusual, and one may prefer to replace it by $\|g\|_{-1,q;D}$ for example, the norm of g in $W_0^{-1,q}(\Omega)$. One can show that this is not possible in general; see [42, Section III.3]. However, the following proposition shows that when $g = u \cdot \nabla\chi = \operatorname{div}(\chi u)$, the situation is different.

Proposition 2.4.3. *Let $\Omega \subset \mathbb{R}^n$ be an exterior domain and let $D \subset \Omega$ be a bounded domain, both with Lipschitz boundary. Let $q \in (1, \infty)$ and $u \in W_{\mathrm{loc}}^{1,q}(\Omega)^n$ with $\operatorname{div} u = 0$ in Ω and $u \cdot n = 0$ on $\partial\Omega$. Further, let $a, b \in \mathbb{R}$ and let $\chi \in C^\infty(\mathbb{R}^n)$ be a smooth function with $\chi \equiv a$ on B^R and $\chi \equiv b$ on B_r for some $R > r > \delta(\Omega^c)$ such that $\mathrm{B}_R \setminus \mathrm{B}_r \subset D$. Then*

$$|u \cdot \nabla\chi|_{-1,q;D}^* \leq C_{13} \|u\|_{-1,q;D}. \tag{2.12}$$

for some constant $C_{13} = C_{13}(n, D, q, \chi) > 0$.

Proof. To show (2.12), for $\psi \in C_0^\infty(\mathbb{R}^n)$ we introduce the notation

$$\psi_D := \frac{1}{|D|} \int_D \psi \, \mathrm{d}x.$$

Set $g := u \cdot \nabla\chi = \operatorname{div}(\chi u)$. By assumption we have $\operatorname{supp} g \subset B_R \setminus B_r \subset D$. Due to $u \cdot \mathrm{n} = 0$ on $\partial\Omega$, the divergence theorem yields

$$\int_{B_R} g\psi \, \mathrm{d}x = \int_{\partial B_R} \chi u \cdot \mathrm{n}\psi \, \mathrm{d}S - \int_{B_R} \chi u \cdot \nabla\psi \, \mathrm{d}x$$

$$= \int_{\partial B_R} \chi u \cdot \mathrm{n}\psi \, \mathrm{d}S - \int_{B_R} \chi u \cdot \nabla(\psi - \psi_D) \, \mathrm{d}x$$

$$= a\psi_D \int_{\partial B_R} u \cdot \mathrm{n} \, \mathrm{d}S + \int_{B_R} g(\psi - \psi_D) \, \mathrm{d}x = \int_{B_R} g(\psi - \psi_D) \, \mathrm{d}x$$

due to $\operatorname{div} u = 0$. Repeating this calculation for B_r, we obtain the analogue identity and conclude

$$\int_D g\psi \, \mathrm{d}x = \int_{B_R} g\psi \, \mathrm{d}x - \int_{B_r} g\psi \, \mathrm{d}x = \int_{B_R} g(\psi - \psi_D) \, \mathrm{d}x - \int_{B_r} g(\psi - \psi_D) \, \mathrm{d}x$$

$$= \int_D g(\psi - \psi_D) \, \mathrm{d}x = \int_D u \cdot \nabla\chi(\psi - \psi_D) \, \mathrm{d}x.$$

Therefore, we have

$$\left| \int_D g\psi \, \mathrm{d}x \right| \le \|u\|_{-1,q;D} \|\nabla\chi(\psi - \psi_D)\|_{1,q';D}$$

with $q' = q/(q-1)$. From $\chi \in C_0^\infty(\mathbb{R}^n)$ and Poincaré's inequality, we further deduce

$$\|(\psi - \psi_D)\nabla\chi_0\|_{1,q';D} \le c_0 \|\psi - \psi_D\|_{1,q';D} \le c_1 \|\nabla(\psi - \psi_D)\|_{q';D} = c_1 \|\nabla\psi\|_{q';D}.$$

Employing both of these estimates in the definition of $|\cdot|_{-1,q';D}^*$, we obtain

$$|u \cdot \nabla\chi|_{-1,q;D}^* = |g|_{-1,q;D}^*$$
$$\le \sup\{\|u\|_{-1,q;D}\|(\psi - \psi_D)\nabla\chi\|_{1,q';D} \mid \psi \in C_0^\infty(\mathbb{R}^3), \ \|\nabla\psi\|_{q';D} = 1\}$$
$$\le c_1 \|u\|_{-1,q;D} \sup\{\|\nabla\psi\|_{q';D} \mid \psi \in C_0^\infty(\mathbb{R}^3), \ \|\nabla\psi\|_{q';D} = 1\}$$
$$= c_1 \|u\|_{-1,q;D},$$

which is (2.12). □

We can now combine the estimates (2.10) and (2.12). Moreover, in the setting of Proposition 2.4.3, assumption (2.9) is satisfied automatically, so that the Bogovskiĭ operator provides a solution to problem (2.8).

Corollary 2.4.4. *In the situation of Proposition 2.4.3, let $\mathfrak{B}$ be the Bogovskiĭ operator on D. Then* $\operatorname{div}\mathfrak{B}(u \cdot \nabla\chi) = u \cdot \nabla\chi$ *and*

$$\|\mathfrak{B}(u \cdot \nabla\chi)\|_{q;D} \leq C_{14}\|u\|_{-1,q;D}. \tag{2.13}$$

Proof. Estimate (2.13) is a direct consequence of (2.10) and (2.12). It remains to verify condition (2.9). As in the previous proof, we define $g := u \cdot \nabla\chi$. By the divergence theorem and due to $u \cdot n = 0$ on $\partial\Omega$, we obtain

$$\int_{B_R} g \, dx = \int_{B_R} \operatorname{div}(\chi u) \, dx = \int_{\partial B_R} \chi u \cdot n \, dS = a \int_{B_R} \operatorname{div} u \, dx = 0.$$

Repeating this calculation for B_r, we see that the corresponding integral vanishes as well. Due to $\operatorname{supp} g \subset B_R \setminus B_r$, we obtain

$$\int_D g \, dx = \int_{B_R \setminus B_r} g \, dx = \int_{B_R} g \, dx - \int_{B_r} g \, dx = 0.$$

Now Theorem 2.4.2 yields $\operatorname{div}\mathfrak{B}g = g$, which completes the proof. $\square$

2.4.3 The Stokes Problem in a Bounded Domain

Here we collect several results concerning the analysis of the (generalized) Stokes resolvent problem

$$\begin{cases} \lambda u - \Delta u + \nabla\mathfrak{p} = f & \text{in } D, \\ \operatorname{div} u = g & \text{in } D, \\ u = 0 & \text{on } \partial D \end{cases} \tag{2.14}$$

in a bounded domain $D \subset \mathbb{R}^n$. First of all, let us recall the following well-posedness result including a resolvent estimate.

Theorem 2.4.5 (Stokes Resolvent Problem). *Let $D \subset \mathbb{R}^n$, $n \geq 2$, be a bounded domain of class C^2. Let $\varepsilon > 0$ and*

$$\lambda \in \{z \in \mathbb{C} \setminus \{0\} \mid |\arg z| < \pi - \varepsilon\} \cup \{0\}.$$

For all $f \in \mathrm{L}^q(D)$ and $g \in \mathrm{W}^{1,q}(D)$ with

$$\int_D g \, dx = 0$$

there exist functions $u \in \mathrm{W}^{2,q}(D)$ and $\mathfrak{p} \in \mathrm{W}^{1,q}(D)$ that satisfy (2.14), and there is a constant $C_{15} = C_{15}(n, D, q, \varepsilon) > 0$ such that

$$\|\lambda u\|_q + \|u\|_{2,q} + \|\nabla \mathfrak{p}\|_q \leq C_{15}\big(\|f\|_q + \|\nabla g\|_q + |\lambda g|^*_{-1,q;D}\big), \tag{2.15}$$

where $|\cdot|^*_{-1,q;D}$ is defined in (2.11). Moreover, if $(u_1, \mathfrak{p}_1) \in \mathrm{W}^{2,q}(D) \times \mathrm{W}^{1,q}(D)$ is another solution to (2.14), then $u = u_1$ and $\mathfrak{p} = \mathfrak{p}_1 + c$ for some constant $c \in \mathbb{R}$.

Proof. See [33] for example. $\qquad\square$

Observe that the constant C_{15} in the resolvent estimate (2.15) is independent of the resolvent parameter λ, which is crucial for the treatment of corresponding time-dependent problems. Moreover, the term $|\lambda g|^*_{-1,q;D}$, which appears on the right-hand side of (2.15), can be estimated with the help of Proposition (2.4.3) if g is of the form discussed there.

In the divergence-free case, that is, for $g = 0$, one can reformulate (2.14) as a resolvent problem of the *Stokes operator* $-\mathcal{P}_\mathrm{H}\Delta$, which is a closed operator on the space of solenoidal vector field $\mathrm{L}^q_\sigma(D)$. The following theorem collects famous results in the Hilbert-space case $q = 2$.

Theorem 2.4.6 (Stokes Operator). *Let $D \subset \mathbb{R}^n$, $n \geq 2$, be a bounded domain of class C^2. Then the Stokes operator, given by*

$$A \colon \mathrm{dom}(A) \subset \mathrm{L}^2_\sigma(D) \to \mathrm{L}^2_\sigma(D), \quad u \mapsto -\mathcal{P}_\mathrm{H}\Delta u, \\ \mathrm{dom}(A) := \mathrm{L}^2_\sigma(D) \cap \mathrm{W}^{1,2}_0(D)^n \cap \mathrm{W}^{2,2}(D)^n, \tag{2.16}$$

is a densely defined closed operator. Moreover, A is invertible and positive self-adjoint. If $f \in \mathrm{L}^2(\Omega)$, then $(u, \mathfrak{p}) \in \mathrm{W}^{2,2}(\Omega) \times \mathrm{W}^{1,2}(\Omega)$ is a solution to

$$\begin{cases} -\Delta u + \nabla \mathfrak{p} = f & \text{in } D, \\ \mathrm{div}\, u = 0 & \text{in } D, \\ u = 0 & \text{on } \partial D \end{cases} \tag{2.17}$$

if and only if $u \in \mathrm{dom}(A)$ with $Au = \mathcal{P}_\mathrm{H}f$.

Proof. See [97, Theorem III.2.1.1] for example. $\qquad\square$

For the moment consider $f \in L^2_\sigma(D)$, and let $(u, \mathfrak{p}) \in W^{2,2}(\Omega) \times W^{1,2}(\Omega)$ be a solution to (2.17). By Theorem 2.4.6 we then have $-\mathcal{P}_H \Delta u = \mathcal{P}_H f = f$, and Theorem 2.4.5 yields the estimate

$$\|u\|_{2,2} \leq C_{15}\|f\|_2 = C_{15}\|\mathcal{P}_H \Delta u\|_2,$$

where the constant C_{15} depends on the domain D, which is a natural phenomenon. The peculiarity of the next lemma is that it yields an estimate of the second derivatives of u by the Stokes operator such that the corresponding constant is independent of the domain D under certain conditions. However, to achieve this, one has to add the term $\|\nabla u\|_2$ on the right-hand side.

Lemma 2.4.7. *Let $D \subset \mathbb{R}^3$ be a bounded domain with C^3-boundary. Every $u \in L^2_\sigma(D) \cap W^{1,2}_0(D) \cap W^{2,2}(D)$ satisfies*

$$\|\nabla^2 u\|_2 \leq C_{16}\big(\|\mathcal{P}_H \Delta u\|_2 + \|\nabla u\|_2\big)$$

for a constant $C_{16} = C_{16}(D) > 0$ that does not depend on the "size" of D but solely on its "regularity". In particular, if $D = \Omega_R$ for an exterior domain Ω with $\partial\Omega \subset B_R$, the constant C_{16} is independent of R and solely depends on Ω.

Proof. See [59, Lemma 1]. □

3 Flow Past a Non-rotating Body

This chapter is concerned with the existence of strong solutions to the Oseen and Navier–Stokes equations, describing viscous incompressible flow past a body, in both the steady-state and the time-periodic case. The main novelty is the following: Though considering these problems in an exterior domain, we derive existence of solutions with velocity field belonging to a *full* Sobolev space, and not only to a homogeneous one. These results are new and were published in [19].

To be more precise, we begin with the linear theory, that is, the study of the time-periodic Oseen problem

$$
\begin{cases}
\partial_t u - \Delta u - \lambda \partial_1 u + \nabla \mathfrak{p} = f & \text{in } \mathbb{T} \times \Omega, \\
\operatorname{div} u = 0 & \text{in } \mathbb{T} \times \Omega, \\
u = 0 & \text{on } \mathbb{T} \times \partial\Omega, \\
\lim_{|x| \to \infty} u(t, x) = 0 & \text{for } t \in \mathbb{T}
\end{cases}
$$

and its steady-state counterpart in an exterior domain $\Omega \subset \mathbb{R}^n$, $n \geq 2$. The Reynolds number $\lambda > 0$ and the time period $\mathcal{T} > 0$ defining the torus group $\mathbb{T} := \mathbb{R}/T\mathbb{Z}$ are fixed and data f is prescribed. The main task is to find suitable functional frameworks that render these problems well posed, that is, such that there exists a unique velocity field u and a pressure $\mathfrak{p}$ in a suitable functional framework that satisfy these equations and obey corresponding *a priori* estimates.

Let us address the time-independent case at first, that is, the steady-state Oseen problem, which has been studied extensively over the last years. There exists a well-established L^q theory for this problem, which was initiated by GALDI [40]. For an overview, we further refer to [42, Chapter VII], the more recent paper [2] and the bibliography there included. Concerning the question of well-posedness, the peculiarity is that for general data $f \in L^q(\Omega)$, the solution space for the velocity field u is not a *full* Sobolev space but an intersection of *homogeneous* Sobolev spaces. One can merely ensure that

$$\nabla^2 v \in L^q(\Omega), \qquad \nabla v \in L^{s_1}(\Omega), \qquad v \in L^{s_2}(\Omega)$$

for *different* values $q < s_1 < s_2$. In this chapter, this problem is resolved by passing to a different functional framework. More precisely, we consider data f that, besides being an element of a Lebesgue space $L^q(\Omega)$, belong to $D_0^{-1,r}(\Omega)$, the dual space of a homogeneous Sobolev space. Under certain conditions on q and r, we show existence of a solution $(u, \mathfrak{p})$ such that the velocity field u belongs to the full space $W^{2,q}(\Omega)$ and satisfies corresponding *a priori* estimates.

Based on this, we analyze the time-periodic Oseen problem by decomposing it into the steady-state Oseen problem and a second *purely periodic* problem. For the latter, the velocity-field solution naturally belongs to the full Sobolev space $W^{1,2,q}(\mathbb{T} \times \Omega)$ for $f \in L^q(\mathbb{T} \times \Omega)$. Therefore, by a combination with the previously collected steady-state results, we then establish well-posedness in a framework where the (time-periodic) velocity field belongs to the full Sobolev space $W^{1,2,q}(\mathbb{T} \times \Omega)$.

Finally, we show existence of both steady-state and time-periodic solutions to the Navier–Stokes problem

$$\begin{cases} \partial_t v - \Delta v - \lambda \partial_1 v + \nabla p + v \cdot \nabla v = f & \text{in } \mathbb{T} \times \Omega, \\ \operatorname{div} v = 0 & \text{in } \mathbb{T} \times \Omega, \\ v = -\lambda e_1 & \text{on } \mathbb{T} \times \partial\Omega, \\ \lim_{|x| \to \infty} v(t, x) = 0 & \text{for } t \in \mathbb{T}, \end{cases}$$

which describes the flow around a body that translates with non-vanishing velocity λe_1, $\lambda > 0$. Assuming that the data f and λ are "sufficiently small", and $n \geq 3$, we employ the developed linear theory and show existence of solutions (v, p) by the contraction mapping principle. This yields solutions where the velocity field v belongs to a full Sobolev space in both the steady-state and the time-periodic case.

In this chapter, we proceed as follows. In Section 3.1 we show well-posedness of the steady-state Oseen problem in a framework where the velocity field belongs to the full Sobolev space $W^{2,q}(\Omega)$. In Section 3.2 we then derive two different well-posedness results for the time-periodic Oseen problem. Finally, Section 3.3 deals with the nonlinear case, that is, the existence of steady-state and time-periodic solutions to the Navier–Stokes equations for "small" data.

3.1 The Steady-State Oseen System

In this section, we address the question of well-posedness of the steady-state Oseen system

$$
\begin{cases}
-\Delta u - \lambda \partial_1 u + \nabla \mathfrak{p} = f & \text{in } \Omega, \\
\operatorname{div} u = 0 & \text{in } \Omega, \\
u = 0 & \text{on } \partial\Omega, \\
\lim_{|x| \to \infty} u(x) = 0,
\end{cases}
\tag{3.1}
$$

where Ω is an exterior domain of $\mathbb{R}^n$ and $f: \Omega \to \mathbb{R}^n$ and $\lambda > 0$ are a given external force and a dimensionless (Reynolds) number, whereas $u: \Omega \to \mathbb{R}^n$ and $\mathfrak{p}: \Omega \to \mathbb{R}$ are unknown velocity and pressure fields, respectively. First of all, we recall well-known results on the existence of strong solutions for data $f \in L^q(\Omega)$ and of weak solutions for data $f \in D_0^{-1,r}(\Omega)$. In both cases one can merely ensure the velocity field u to belong to an intersection of homogeneous Sobolev spaces but not to a classical one. Motivated by this observation, we then derive a solution theory such that the velocity field u belongs to a classical Sobolev space $W^{2,q}(\Omega)$ for suitable $q \in (1, \infty)$. To this end, we consider data $f \in L^q(\Omega) \cap D_0^{-1,r}(\Omega)$ and establish existence of a unique solution together with suitable *a priori* estimates.

3.1.1 Strong Solutions in Homogeneous Sobolev Spaces

Here we recall the classical well-posedness result for the steady-state Oseen problem (3.1) for data $f \in L^q(\Omega)$. For this purpose, we introduce the following function space, which characterizes the velocity field u. For $q \in (1, \frac{n+1}{2})$ and $\lambda > 0$ we define

$$X_\lambda^q(\Omega) := \left\{ u \in D^{2,q}(\Omega)^n \cap D^{1,s_1}(\Omega)^n \cap L^{s_2}(\Omega)^n \mid \partial_1 u \in L^q(\Omega)^n \right\}, \quad (3.2)$$

which we equip with the norm

$$\|u\|_{X_\lambda^q(\Omega)} := |u|_{2,q} + \lambda \|\partial_1 u\|_q + \lambda^{\frac{1}{n+1}} |u|_{1,s_1} + \lambda^{\frac{2}{n+1}} \|u\|_{s_2},$$

where

$$s_1 := \frac{(n+1)q}{n+1-q}, \qquad s_2 := \frac{(n+1)q}{n+1-2q}. \quad (3.3)$$

The following result is well known.

Theorem 3.1.1. *Let $\Omega \subset \mathbb{R}^n$, $n \geq 2$, be an exterior domain with C^2 boundary. Let $\lambda > 0$ and $q \in (1, \frac{n+1}{2})$. For every $f \in L^q(\Omega)$ there exists a solution $(u, \mathfrak{p}) \in X_\lambda^q(\Omega) \times D^{1,q}(\Omega)$ to (3.1), which further satisfies the estimate*

$$\|u\|_{X_\lambda^q(\Omega)} + |\mathfrak{p}|_{1,q} \leq C_{17} \|f\|_q \quad (3.4)$$

for a constant $C_{17} = C_{17}(n, \Omega, q, \lambda) > 0$. However, if $q \in (1, \frac{n}{2})$ and $0 < \lambda \leq \lambda_0$ for some $\lambda_0 > 0$, then $C_{17} = C_{17}(n, \Omega, q, \lambda_0)$, that is, C_{17} is independent of λ. Moreover, if $(\tilde{u}, \tilde{\mathfrak{p}}) \in X_\lambda^q(\Omega) \times D^{1,q}(\Omega)$ is another solution to (3.1), then $u = \tilde{u}$ and $\mathfrak{p} = \tilde{\mathfrak{p}} + c$ for some constant $c \in \mathbb{R}$.

Proof. This result was proved in [40]. See also [42, Theorem VII.7.1]. $\quad\square$

The previous theorem does not treat the case $f \in L^q(\Omega)$ for $q \geq \frac{n+1}{2}$. However, it is possible to show existence of a unique solution in an appropriate quotient space, which satisfies an *a priori* estimate corresponding to (3.4); see [42, Remark VII.7.1] for example.

To obtain a well-posedness result in a different functional framework, one main ingredient in our approach is the following theorem that provides us with an *a priori* estimate for a solution to the Oseen problem

$$\begin{cases} -\Delta u - \lambda \partial_1 u + \nabla \mathfrak{p} = f & \text{in } \mathbb{R}^n, \\ \operatorname{div} u = 0 & \text{in } \mathbb{R}^n \end{cases} \quad (3.5)$$

in the whole space.

Theorem 3.1.2. *Let $n \geq 2$, $q \in (1, \infty)$ and $\lambda > 0$. For every $f \in L^q(\mathbb{R}^n)$ there exists a solution $(u, \mathfrak{p}) \in W^{2,q}_{loc}(\mathbb{R}^n)^n \times W^{1,q}_{loc}(\mathbb{R}^n)$ to (3.5) that satisfies*

$$|u|_{2,q} + \lambda \|\partial_1 u\|_q + |\mathfrak{p}|_{1,q} \leq C_{18} \|f\|_q \tag{3.6}$$

for some constant $C_{18} = C_{18}(n, q) > 0$ independent of λ.

Proof. See [42, Theorem VII.4.1]. □

Note that Theorem 3.1.2 tells nothing about uniqueness of the solution. Clearly, if one modifies u and $\mathfrak{p}$ by additive constants, they still are solutions to (3.5). This is not the only possible modification. However, in the next lemma we show that we have uniqueness up to polynomials. Note that we also allow the case $\lambda = 0$.

Lemma 3.1.3. *Let $n \geq 2$, $\lambda \in \mathbb{R}$ and $(u, \mathfrak{p}) \in \mathscr{S}'(\mathbb{R}^n)^{n+1}$ satisfy (3.5) with data $f = 0$. Then u_j, $j = 1, \ldots, n$, and $\mathfrak{p}$ are polynomials.*

Proof. Computing the divergence of $(3.5)_1$, we obtain $\Delta \mathfrak{p} = 0$, which yields $-|\xi|^2 \widehat{\mathfrak{p}} = 0$ with $\widehat{\mathfrak{p}} = \mathscr{F}_{\mathbb{R}^n}[\mathfrak{p}]$. Hence, we have $\operatorname{supp} \widehat{\mathfrak{p}} \subset \{0\}$, so that $\mathfrak{p}$ is a polynomial. Now an application of the Fourier transform $\mathscr{F}_{\mathbb{R}^n}$ to $(3.5)_1$ results in $(|\xi|^2 - i\lambda\xi_1)\widehat{u} = -i\xi\widehat{\mathfrak{p}}$ with $\widehat{u} = \mathscr{F}_{\mathbb{R}^n}[u]$, so that

$$\operatorname{supp}\left[(|\xi|^2 - i\lambda\xi_1)\widehat{u}\right] = \operatorname{supp}\left[-i\xi\widehat{\mathfrak{p}}\right] \subset \{0\}.$$

Since the only zero of $\xi \mapsto (|\xi|^2 - i\lambda\xi_1)$ is $\xi = 0$, we conclude $\operatorname{supp} \widehat{u} \subset \{0\}$, and u is a polynomial in each component. □

3.1.2 Weak Solutions in Homogeneoues Sobolev Spaces

After having prepared the required statements on strong solutions, let us now collect some well-known results about weak solutions to the steady-state Oseen problem (3.1). First of all, let us recall the notion of weak solutions.

Definition 3.1.4. Let $\Omega \subset \mathbb{R}^n$, $n \geq 2$, be an exterior domain, and let $f \in \mathcal{D}'(\Omega)^n$ be a distribution. A function $u \in D_0^{1,r}(\Omega)^n$ with $r \in (1, \infty)$ is called *weak solution* to (3.1) if it satisfies $\operatorname{div} u = 0$, and

$$\int_\Omega \nabla u : \nabla \varphi - \lambda \partial_1 u \cdot \varphi \, dx = \langle f, \varphi \rangle \tag{3.7}$$

for all $\varphi \in C_{0,\sigma}^\infty(\Omega)$. Moreover, we call a function $\mathfrak{p} \in L_{loc}^1(\Omega)$ *a pressure associated to* u if $(u, \mathfrak{p})$ satisfies $(3.1)_1$ in the sense of distributions, that is,

$$\int_\Omega \nabla u : \nabla \psi - \lambda \partial_1 u \cdot \psi \, dx = \langle f, \varphi \rangle + \int_\Omega \mathfrak{p} \operatorname{div} \psi \, dx \tag{3.8}$$

for all $\psi \in C_0^\infty(\Omega)^n$.

The next lemma provides sufficient conditions on the right-hand side f in order to find an associated pressure $\mathfrak{p}$ for a weak solution u.

Lemma 3.1.5. *Let $\Omega \subset \mathbb{R}^n$, $n \geq 2$, be a locally Lipschitz exterior domain, let $r \in (1, \infty)$ and $f \in W_0^{-1,r}(\Omega_R)^n$ for any $R > \delta(\Omega^c)$. Then, for every weak solution $u \in D_0^{1,r}(\Omega)^n$ to (3.1) there exists an associated pressure field $\mathfrak{p}$ with $\mathfrak{p} \in L^r(\Omega_R)$ for all $R > \delta(\Omega^c)$. If $f \in D_0^{-1,r}(\Omega)^n$ for $r > n/(n-1)$, one can choose $\mathfrak{p}$ such that $\mathfrak{p} \in L^r(\Omega)$.*

Proof. See [42, Lemma VII.1.1 and Theorem VII.7.3]. □

Note that Lemma 3.1.5 only yields a pressure $\mathfrak{p} \in L^r(\Omega)$ if $r > n/(n-1)$. It is an open question whether or not this global integrability can be achieved if $r \in (1, n/(n-1)]$; see also [42, Remark VII.7.2].

The existence of weak solutions $u \in D_0^{1,r}(\Omega)$ for $f \in D_0^{-1,r}(\Omega)$ together with corresponding *a priori* estimates has first been shown by GALDI [40] in the case $r \in (\frac{n}{n-1}, n+1)$. Later, AMROUCHE and RAZAFISON [3] extended this result in the three-dimensional case to $r \in (\frac{4}{3}, 4)$, and KIM and KIM [64] generalized it in $n \geq 2$ dimensions and showed existence for $f \in D_0^{-1,r}(\Omega)^n$ for any $r \in (\frac{n+1}{n}, n+1)$. The following theorem collects their results.

Theorem 3.1.6. *Let $\Omega \subset \mathbb{R}^n$, $n \geq 2$, be an exterior domain with C^2-boundary. Let $r \in (\frac{n+1}{n}, n+1)$, $0 < \lambda \leq \lambda_0$ and $f \in D_0^{-1,r}(\Omega)^n$. Define $s := \frac{(n+1)r}{n+1-r}$. Then there exists a unique weak solution $u \in D_0^{1,r}(\Omega)^n \cap L^s(\Omega)^n$ to (3.1), which satisfies the estimate*

$$\|\nabla u\|_r + \lambda^{\frac{1+\delta}{n+1}} \|u\|_s \leq C_{19} \lambda^{-\frac{M}{n+1}} |f|_{-1,r} \tag{3.9}$$

for some constant $C_{19} = C_{19}(n, r, \Omega, \lambda_0) > 0$, where

$$M = \begin{cases} 2 & if \ \frac{n+1}{n} < r \leq \frac{n}{n-1}, \\ 0 & if \ \frac{n}{n-1} < r < n, \\ 1 & if \ n \leq r < n+1, \end{cases} \qquad \delta = \begin{cases} 1 & if \ n = r = 2, \\ 0 & else. \end{cases} \tag{3.10}$$

Furthermore, there exists an associated pressure $\mathfrak{p}$ *such that* $\mathfrak{p} \in L^r(\Omega_R)$ *and* $\partial_1 u \in W_0^{-1,r}(\Omega_R)^n$ *for any* $R > \delta(\Omega^c)$. *Provided that*

$$\int_{\Omega_R} \mathfrak{p} \, dx = 0,$$

these functions further obey the estimate

$$\lambda \|\partial_1 u\|_{-1,r;\Omega_R} + \|\mathfrak{p}\|_{r;\Omega_R} \leq C_{20} \lambda^{-\frac{M}{n+1}} |f|_{-1,r} \tag{3.11}$$

for a constant $C_{20} = C_{20}(n, r, \Omega, \lambda_0, R) > 0$.

Additionally, if $r \in (\frac{n}{n-1}, n)$, *we can choose* $\mathfrak{p}$ *such that* $\mathfrak{p} \in L^r(\Omega)$. *Then* $\partial_1 u \in D_0^{-1,r}(\Omega)^n$ *and the estimate*

$$\|\mathfrak{p}\|_r + \lambda |\partial_1 u|_{-1,r} \leq C_{21} |f|_{-1,r} \tag{3.12}$$

holds for some constant $C_{21} = C_{21}(n, r, \Omega, \lambda_0) > 0$.

Proof. The existence of a unique weak solution $u \in L^s(\Omega)^n \cap D_0^{1,r}(\Omega)^n$, which obeys estimate (3.9), was established in [64, Theorem 2.2]. Due to the embedding $D_0^{-1,r}(\Omega) \hookrightarrow W_0^{-1,r}(\Omega_R)$, by Lemma 3.1.5 there exists an associated pressure $\mathfrak{p}$ with $\mathfrak{p} \in L^r(\Omega_R)$ for any $R > \delta(\Omega^c)$. Moreover, for $\psi \in C_0^\infty(\Omega_R)^n$ we thus have

$$-\lambda \int_\Omega \partial_1 u \cdot \psi \, dx = \langle f, \psi \rangle + \int_\Omega \left(\mathfrak{p} \operatorname{div} \psi - \nabla u : \nabla \psi \right) dx$$

$$\leq \left(|f|_{-1,r} + \|\mathfrak{p}\|_{r;\Omega_R} + \|\nabla u\|_r \right) \|\nabla \psi\|_{r'},$$

where $r' = r/(r-1)$, which implies $\partial_1 u \in W_0^{-1,r}(\Omega_R)$ and

$$\lambda \|\partial_1 u\|_{-1,r;\Omega_R} \leq c_0 \left(|f|_{-1,r} + \|\mathfrak{p}\|_{r;\Omega_R} + \|\nabla u\|_r \right). \tag{3.13}$$

In particular, since $C_0^\infty(\Omega_R)$ is dense in $W_0^{1,r'}(\Omega_R)$, the identity (3.8) holds for all $\psi \in W_0^{1,r'}(\Omega_R)^n$. To derive an estimate for $\mathfrak{p}$, we fix $R > \delta(\Omega^c)$. By adding a suitable constant to $\mathfrak{p}$, we may assume that its mean value on Ω_R vanishes. Now let $\psi \in W_0^{1,r'}(\Omega_R)^n$ be a solution to the problem

$$\operatorname{div} \psi = |\mathfrak{p}|^{r-2} \mathfrak{p} - \frac{1}{|\Omega_R|} \int_{\Omega_R} |\mathfrak{p}|^{r-2} \mathfrak{p} \, dx =: g \qquad \text{in } \Omega_R,$$

which exists by Theorem 2.4.2 since g has vanishing mean value and satisfies $g \in L^{r'}(\Omega_R)$. Moreover, we have

$$\|\psi\|_{1,r';\Omega_R} \leq c_1 \|g\|_{r';\Omega_R} \leq c_2 \|\mathfrak{p}\|_{r;\Omega_R}^{r-1}.$$

Inserting this function ψ into (3.8), since the mean value of $\mathfrak{p}$ vanishes, we deduce

$$\|\mathfrak{p}\|^r_{r;\Omega_R} = \int_{\Omega_R} \mathfrak{p} \operatorname{div} \psi \, dx + \int_{\Omega_R} \mathfrak{p} \, dx \, \frac{1}{|\Omega_R|} \int_{\Omega_R} |\mathfrak{p}|^{r-2} \mathfrak{p} \, dx$$

$$= \int_{\Omega_R} \nabla u : \nabla \psi - \lambda \partial_1 u \cdot \psi \, dx - \langle f, \psi \rangle$$

$$\leq c_3 (1 + \lambda_0) \big(\|\nabla u\|_r + |f|_{-1,r} \big) \|\psi\|_{1,r';\Omega_R}$$

$$\leq c_4 (1 + \lambda_0) \big(\|\nabla u\|_r + |f|_{-1,r} \big) \|\mathfrak{p}\|^{r-1}_{r;\Omega_R}.$$

Combining this estimate, (3.13) and (3.9), we arrive at (3.11).

Now let $n/(n-1) < r < n$. Then Lemma 3.1.5 yields the existence of a pressure $\mathfrak{p} \in L^r(\Omega)$, and by [42, Theorem VII.7.2] we have the estimate

$$\|\mathfrak{p}\|_r \leq c_5 |f|_{-1,r}.$$

Arguing as above, we now obtain $\partial_1 u \in D_0^{-1,r}(\Omega)$ and

$$\lambda |\partial_1 u|_{-1,r} \leq c_6 \big(|f|_{-1,r} + \|\mathfrak{p}\|_r + \|\nabla u\|_r \big).$$

Since $M = 0$, a combination of these estimates with (3.9) finally implies (3.12). $\qquad \square$

We further recall the following local regularity result for weak solutions.

Theorem 3.1.7. *Let Ω be an exterior domain of class C^2, $r \in (1, \infty)$, and $f \in C_0^\infty(\Omega)^n$. Let $u \in W^{1,r}_{\mathrm{loc}}(\Omega)^n$, $\mathfrak{p} \in L^r_{\mathrm{loc}}(\Omega)$ such that $\operatorname{div} u = 0$, $u|_{\partial\Omega} = 0$ and (3.7) is satisfied for all $\varphi \in C^\infty_{0,\sigma}(\Omega)$. Then $u \in C^\infty(\Omega) \cap W^{2,r}_{\mathrm{loc}}(\overline{\Omega})$ and the associated pressure $\mathfrak{p}$ satisfies $\mathfrak{p} \in C^\infty(\Omega) \cap W^{1,r}_{\mathrm{loc}}(\overline{\Omega})$.*

Proof. See [42, Theorem VII.1.1]. $\qquad \square$

In order to obtain uniqueness for solutions to (3.1) in an exterior domain, we cannot directly proceed as in Lemma 3.1.3. However, a standard cut-off argument combined with the following decay property for weak solutions turns out to be a useful method to exploit Lemma 3.1.3 also in exterior domains. We employ this in the proof of Lemma 3.1.9.

Theorem 3.1.8. *Let $\Omega \subset \mathbb{R}^n$, $n \geq 2$, be an exterior domain and u be a weak solution to (3.1) with $u \in L^s(\Omega^R)$ for some $s \in (1, \infty)$ and some $R > \delta(\Omega^c)$. Moreover, let $f \in W^{m,r}(\Omega)$ for some $m \in \mathbb{N}_0$, $r \in (n/2, \infty)$. Then*

$$\forall \alpha \in \mathbb{N}_0^n, \ 0 \leq |\alpha| \leq m : \quad \lim_{|x| \to \infty} D^\alpha u(x) = 0.$$

Proof. See [42, Theorem VII.6.1]. $\qquad \square$

3.1.3 Strong Solutions in Full Sobolev Spaces

In order to find an appropriate function space such that the velocity-field solution u to (3.1) is an element of the *inhomogeneous* Sobolev space $W^{2,q}(\Omega)$, we combine the theory for strong and weak solutions from the previous subsections. The following lemma provides an L^q estimate for the solution with error terms on the right-hand side, which are omitted in the proof of the subsequent main theorem. For the proof we employ a classical cut-off procedure.

Lemma 3.1.9. *Let $\Omega \subset \mathbb{R}^n$, $n \geq 2$, be an exterior domain of class C^2, and $r, s \in (1, \infty)$, $f \in C_0^\infty(\Omega)^n$. Let $u \in D^{1,r}(\Omega)^n \cap L^s(\Omega)^n$ be a weak solution to (3.1), and let $\mathfrak{p} \in L_{loc}^r(\Omega)$ be an associated pressure. Then, $u \in D^{2,q}(\Omega)^n$, $\partial_1 u \in L^q(\Omega)^n$, and $\mathfrak{p} \in D^{1,q}(\Omega)$ for all $q \in (1, \infty)$, and for each $R > \delta(\Omega^c)$ there exists $C_{22} = C_{22}(n, q, \Omega, R) > 0$ such that*

$$|u|_{2,q} + \lambda\|\partial_1 u\|_q + |\mathfrak{p}|_{1,q} \leq C_{22}(1 + \lambda^2)\big(\|f\|_q + \|u\|_{1,q;\Omega_R} + \|\mathfrak{p}\|_{q;\Omega_R}\big). \quad (3.14)$$

Proof. By Theorem 3.1.7, we have $u \in W_{loc}^{2,q}(\overline{\Omega}) \cap C^\infty(\Omega)$ and $\mathfrak{p} \in W_{loc}^{1,q}(\overline{\Omega}) \cap C^\infty(\Omega)$ for all $q \in (1, r]$. Let $0 < R_0 < R_1 < R$ such that $\partial B_{R_0} \subset \Omega$, and let $\chi \in C_0^\infty(\mathbb{R}^n)$ with $\chi \equiv 1$ on B_{R_0} and $\chi \equiv 0$ on B^{R_1}. We set $v := (1 - \chi)u + \mathfrak{B}(u \cdot \nabla \chi)$ and $p := (1 - \chi)\mathfrak{p}$, where $\mathfrak{B}$ denotes the Bogovskiĭ operator; see Theorem 2.4.2. Then $v \in W_{loc}^{2,q}(\mathbb{R}^n) \cap D^{1,r}(\mathbb{R}^n) \cap L^s(\mathbb{R}^n)$ and $p \in W_{loc}^{1,q}(\mathbb{R}^n)$ satisfy

$$\begin{cases} -\Delta v - \lambda\partial_1 v + \nabla p = F & \text{in } \mathbb{R}^n, \\ \operatorname{div} v = 0 & \text{in } \mathbb{R}^n \end{cases} \quad (3.15)$$

with

$$F := (1 - \chi)f + 2\nabla\chi \cdot \nabla u + \Delta\chi u + \lambda\partial_1\chi u - \mathfrak{p}\nabla\chi + \big[-\Delta - \lambda\partial_1\big]\mathfrak{B}(u \cdot \nabla\chi).$$

Moreover, we have $F \in L^q(\Omega)$ with

$$\begin{aligned} \|F\|_q &\leq c_0\big(\|f\|_q + (1 + \lambda)\|u\|_{1,q;\Omega_{R_1}} + \|\mathfrak{p}\|_{q;\Omega_{R_1}} + (1 + \lambda)\|\mathfrak{B}(u \cdot \nabla\chi)\|_{2,q}\big) \\ &\leq c_1\big(\|f\|_q + (1 + \lambda)\|u\|_{1,q;\Omega_{R_1}} + \|\mathfrak{p}\|_{q;\Omega_{R_1}}\big) \end{aligned}$$

by Theorem 2.4.2. From Theorem 3.1.2 we thus conclude the existence of a solution $(v_1, p_1) \in W_{loc}^{2,q}(\mathbb{R}^n) \times W_{loc}^{1,q}(\mathbb{R}^n)$ to (3.15) that satisfies estimate

$$|v_1|_{2,q} + \lambda\|\partial_1 v_1\|_q + |p_1|_{1,q} \leq C_{18}\|F\|_q. \quad (3.16)$$

Now set $w := v - v_1$ and $\mathfrak{q} := p - p_1$. Then $(w, \mathfrak{q})$ is a solution to the homogeneous system (3.5) with $f = 0$. Therefore, $w = v - v_1$ is a polynomial by Lemma 3.1.3. Moreover, from Theorem 3.1.8 and $f \in C_0^\infty(\Omega)$ we conclude

$$\lim_{|x| \to \infty} D^\alpha v(x) = \lim_{|x| \to \infty} D^\alpha u(x) = 0$$

for all $\alpha \in \mathbb{N}_0^n$. For $|\alpha| = 2$ we further have $D^\alpha v_1 \in L^q(\Omega)$, so that the polynomial $D^\alpha w = D^\alpha v - D^\alpha v_1$ must be zero, that is, $D^\alpha w = 0$ for $|\alpha| = 2$. In the same way we conclude $\partial_1 w = 0$ and, since both v and v_1 are solutions to (3.15), also $\nabla \mathfrak{q} = 0$. Hence, we can replace (v_1, p_1) with (v, p) in estimate (3.16). Since $u = v$ and $\mathfrak{p} = p$ on Ω^{R_1}, we thus have

$$|u|_{2,q;\Omega^{R_1}} + \lambda \|\partial_1 u\|_{q;\Omega^{R_1}} + |\mathfrak{p}|_{1,q;\Omega^{R_1}} \leq |v|_{2,q} + \lambda \|\partial_1 v\|_q + |p|_{1,q}$$
$$\leq c_2 \|F\|_q \leq c_3 \big(\|f\|_q + (1+\lambda)\|u\|_{1,q;\Omega_{R_1}} + \|\mathfrak{p}\|_{q;\Omega_{R_1}} \big). \tag{3.17}$$

To derive the estimate near the boundary, we use another cut-off function $\chi_1 \in C_0^\infty(\mathbb{R}^n)$ with $\chi_1 \equiv 1$ on B_{R_1} and $\chi_1 \equiv 0$ on B^R, and we set $w := \chi_1 u$ and $\mathfrak{q} := \chi_1 \mathfrak{p}$. Then $(w, \mathfrak{q}) \in W^{2,q}(\Omega) \times W^{1,q}(\Omega)$ is a solution to the Stokes problem

$$\begin{cases} -\Delta w + \nabla \mathfrak{q} = \chi_1 f - 2\nabla \chi_1 \cdot \nabla u - \Delta \chi_1 u + \chi_1 \lambda \partial_1 u + \mathfrak{p} \nabla \chi_1 & \text{in } \Omega_R, \\ \operatorname{div} w = u \cdot \nabla \chi_1 & \text{in } \Omega_R, \\ w = 0 & \text{on } \partial \Omega_R. \end{cases}$$

From Theorem 2.4.5 we conclude that $(w, \mathfrak{q})$ is subject to the estimate

$$\|w\|_{2,q} + \|\nabla \mathfrak{q}\|_q \leq C_{15}\big(\|f\|_q + (1+\lambda)\|u\|_{1,q;\Omega_R} + \|\mathfrak{p}\|_{q;\Omega_R} \big).$$

Since $u = w$ and $\mathfrak{p} = \mathfrak{q}$ on Ω_{R_1}, a combination of this estimate with (3.17) yields (3.14) for all $q \in (1, r]$.

Using $u \in D^{2,r}(\Omega)$ and $\mathfrak{p} \in D^{1,r}(\Omega)$ and Sobolev embeddings, we now obtain $u \in W_{loc}^{1,r_1}(\Omega)$ and $\mathfrak{p} \in L_{loc}^{r_1}(\Omega)$ for $r_1 = nr/(n-1) > r$, and we can repeat the above argument with r replaced by r_1. By classical bootstrap argument, we thus obtain estimate (3.14) for all $q \in (1, \infty)$. $\square$

After these preparations, we can now establish the following theorem that gives sufficient conditions for the right-hand side f such that the velocity field u is an element of $W^{2,q}(\Omega)$.

Theorem 3.1.10. *Let $\Omega \subset \mathbb{R}^n$, $n \geq 2$, be an exterior domain of class C^2, and let $q \in (1, \infty)$, $r \in \left(\frac{n+1}{n}, n+1\right)$, $0 < \lambda \leq \lambda_0$. Set $s := \frac{(n+1)r}{n+1-r}$. Then, for every $f \in L^q(\Omega)^n \cap D_0^{-1,r}(\Omega)^n$ there exists a solution*

$$u \in D^{2,q}(\Omega)^n \cap D^{1,r}(\Omega)^n \cap L^s(\Omega)^n, \qquad \mathfrak{p} \in D^{1,q}(\Omega)$$

to (3.1). *This solution obeys the estimates*

$$|u|_{1,r} + \lambda^{\frac{1+\delta}{n+1}}\|u\|_s \leq C_{23}\lambda^{-\frac{M}{n+1}}|f|_{-1,r}, \tag{3.18}$$

$$|u|_{2,q} + \lambda\|\partial_1 u\|_q + |\mathfrak{p}|_{1,q} \leq C_{23}\left(\|f\|_q + \lambda^{-\frac{M}{n+1}}|f|_{-1,r}\right) \tag{3.19}$$

for some constant $C_{23} = C_{23}(n,q,r,\Omega,\lambda_0) > 0$, *where*

$$M = \begin{cases} 2 & \text{if } \frac{n+1}{n} < r \leq \frac{n}{n-1}, \\ 0 & \text{if } \frac{n}{n-1} < r < n, \\ 1 & \text{if } n \leq r < n+1, \end{cases} \qquad \delta = \begin{cases} 1 & \text{if } n = r = 2, \\ 0 & \text{else.} \end{cases} \tag{3.20}$$

In particular, if $s \leq q$, *then* $u \in W^{2,q}(\Omega)^n$ *and*

$$\lambda^{\frac{(1+\delta)\theta}{n+1}}\|u\|_q + \lambda^{\frac{(1+\delta)\theta}{2(n+1)}}|u|_{1,q} + |u|_{2,q} \leq C_{23}\left(\|f\|_q + \lambda^{-\frac{M}{n+1}}|f|_{-1,r}\right) \tag{3.21}$$

where

$$\theta := \frac{2qs}{n(q-s)+2qs} \in [0,1]. \tag{3.22}$$

Moreover, if $(u_1,\mathfrak{p}_1)$ *is another solution to* (3.1) *that belongs to the same function class as* $(u,\mathfrak{p})$, *then* $u = u_1$ *and* $\mathfrak{p} = \mathfrak{p}_1 + c$ *for some constant* $c \in \mathbb{R}$.

Additionally, if $r \in (\frac{n}{n-1}, n)$, *one can choose* $\mathfrak{p}$ *such that* $\mathfrak{p} \in L^r(\Omega)$. *Moreover, this implies* $\partial_1 u \in D^{-1,r}(\Omega)$ *and the estimate*

$$\|\mathfrak{p}\|_r + \lambda|\partial_1 u|_{-1,r} \leq C_{24}|f|_{-1,r} \tag{3.23}$$

for some constant $C_{24} = C_{24}(n,r,\Omega,\lambda_0) > 0$.

Proof. For the moment, consider $f \in C_0^\infty(\Omega)$, and let $(u,\mathfrak{p})$ be the unique weak solution to (3.1) satisfying (3.18), which exists due to Theorem 3.1.6. This further yields (3.23) if $n/(n-1) < r < n$. By Lemma 3.1.9, from $f \in C_0^\infty(\Omega)^n$ we conclude $u \in D^{2,q}(\Omega)^n$, $\partial_1 u \in L^q(\Omega)^n$, $\mathfrak{p} \in D^{1,q}(\Omega)$ and the validity of (3.14). Next, let us remove the norms of u and $\mathfrak{p}$ on the right-hand side of (3.14). Corollary 2.3.6 yields the estimates

$$\|u\|_{q;\Omega_R} \leq c_0\left(\|u\|_s + |u|_{1,r}\right) + \varepsilon|u|_{2,q},$$
$$\|\nabla u\|_{q;\Omega_R} \leq c_1|u|_{1,r} + \varepsilon|u|_{2,q},$$
$$\|\mathfrak{p}\|_{q;\Omega_R} \leq c_2\|\mathfrak{p}\|_{r;\Omega_R} + \varepsilon|\mathfrak{p}|_{1,q}.$$

Choosing $\varepsilon > 0$ sufficiently small and combining these with the estimate (3.14), we arrive at

$$|u|_{2,q} + \lambda\|\partial_1 u\|_q + |\mathfrak{p}|_{1,q} \le c_3(1+\lambda^2)\big(\|f\|_q + \|u\|_s + |u|_{1,r} + \|\mathfrak{p}\|_{r;\Omega_R}\big).$$

From (3.18) and $\lambda \le \lambda_0$ we now conclude (3.19) in the case $f \in C_0^\infty(\Omega)$.

Now consider general $f \in L^q(\Omega) \cap D_0^{-1,r}(\Omega)$. By Proposition 2.3.1, there exists a sequence $(f_j) \subset C_0^\infty(\Omega)$ that converges to f. Let $(u_j, \mathfrak{p}_j)$ be the unique solution to (3.1) with right-hand side f_j that satisfies (3.18) and (3.19). Then the differences $(u_j - u_k, \mathfrak{p}_j - \mathfrak{p}_k)$ satisfy (3.1) with right-hand side $f_j - f_k$ and corresponding estimates. Since (f_j) is a Cauchy sequence in $L^q(\Omega) \cap D_0^{-1,r}(\Omega)$, the solutions $(u_j, \mathfrak{p}_j)$ constitute a Cauchy sequence with respect to the norms on the right-hand side of (3.18) and (3.19). Thus, there exists a limit $(u, \mathfrak{p})$, which satisfies (3.18) and (3.19) and is a solution to (3.1) with right-hand side f. In the same way we see that (3.23) is satisfied if $n/(n-1) < r < n$.

Moreover, for $s \le q$ we can employ the Gagliardo–Nirenberg inequality (Theorem 2.3.7) with $\theta \in [0,1]$ given in (3.22) to obtain

$$\|u\|_q \le c_4\|u\|_s^\theta|u|_{2,q}^{1-\theta} \le c_5\lambda^{-\frac{(1+\delta)\theta}{n+1}}\big(\|f\|_q + \lambda^{-\frac{M}{n+1}}|f|_{-1,r}\big)$$

and

$$|u|_{1,q} \le c_6\|u\|_q^{1/2}|u|_{2,q}^{1/2} \le c_7\lambda^{-\frac{(1+\delta)\theta}{2(n+1)}}\big(\|f\|_q + \lambda^{-\frac{M}{n+1}}|f|_{-1,r}\big),$$

where we used (3.18) and (3.19). This shows $u \in W^{2,q}(\Omega)$ together with estimate (3.21) and completes the proof. $\qquad\square$

Remark 3.1.11. Note that the condition $s \le q$ is equivalent to $\frac{1}{q} \le \frac{1}{r} - \frac{1}{n+1}$. Therefore, the assumption $r > \frac{n+1}{n}$ in Theorem 3.1.10 implies the necessary condition $q > \frac{n+1}{n-1}$. Consequently, for any $q \in (\frac{n+1}{n-1}, \infty)$ one can find suitable r such that for any $f \in L^q(\Omega)^n \cap D_0^{-1,r}(\Omega)^n$ there exists a solution with velocity field u in the inhomogeneous Sobolev space $W^{2,q}(\Omega)^n$.

3.2 The Time-Periodic Oseen System

After having dealt with the steady-state Oseen problem (3.1) in an exterior domain $\Omega \subset \mathbb{R}^n$, $n \ge 2$, we now turn to the time-periodic Oseen problem

$$\begin{cases} \partial_t u - \Delta u - \lambda\partial_1 u + \nabla\mathfrak{p} = f & \text{in } \mathbb{T}\times\Omega, \\ \operatorname{div} u = 0 & \text{in } \mathbb{T}\times\Omega, \\ u = 0 & \text{on } \mathbb{T}\times\partial\Omega, \\ \lim_{|x|\to\infty} u(t,x) = 0 & \text{for } t \in \mathbb{T}, \end{cases} \qquad (3.24)$$

with $f: \mathbb{T} \times \Omega \to \mathbb{R}^n$ a given time-periodic external force and $u: \mathbb{T} \times \Omega \to \mathbb{R}^n$ and $\mathfrak{p}: \mathbb{T} \times \Omega \to \mathbb{R}$ the unknown time-periodic velocity and pressure fields. The main idea of our approach is to first single out the steady-state part $(\mathcal{P}u, \mathcal{P}\mathfrak{p})$ of the solution $(u, \mathfrak{p})$, which can be dealt with by applying the theory from the previous section. It thus remains to examine the remaining purely periodic part $(\mathcal{P}_\perp u, \mathcal{P}_\perp \mathfrak{p})$, which, as it turns out, has better functional analytic properties. More precisely, in contrast to the observation for the steady-state problem, the velocity field $\mathcal{P}_\perp u$ directly belongs to the full Sobolev space $W^{1,2,q}(\mathbb{T} \times \Omega)$ for any right-hand side $\mathcal{P}_\perp f \in L^q(\mathbb{T} \times \Omega)$. Subsequently, we combine the results for the steady-state problem and the purely periodic problem to obtain a solution theory for the full time-periodic Oseen problem (3.24).

3.2.1 The Purely Periodic Problem

Here we study well-posedness of the time-periodic Oseen problem (3.24) when the right-hand side f is purely periodic, that is, $\mathcal{P}f = 0$. We obtain that for any $f \in L^q_\perp(\mathbb{T} \times \Omega)$, the solution $(u, \mathfrak{p})$ is also purely periodic, and the velocity field u belongs to the full Sobolev space $W^{1,2,q}(\mathbb{T} \times \Omega)$. This may be seen as the main difference in comparison to the steady-state problem (3.1), for which we stated different well-posedness results for strong solutions, namely Theorem 3.1.1 and Theorem 3.1.10.

Theorem 3.2.1. *Let $\Omega \subset \mathbb{R}^n$, $n \geq 2$, be an exterior domain of class C^2, $q \in (1, \infty)$ and $0 \leq \lambda < \lambda_0$. For any $f \in L^q_\perp(\mathbb{T} \times \Omega)^n$ there is a solution*

$$(u, \mathfrak{p}) \in W^{1,2,q}_\perp(\mathbb{T} \times \Omega)^n \times L^q_\perp(\mathbb{T}; D^{1,q}(\Omega))$$

to (3.24), which satisfies

$$\|u\|_{1,2,q} + \|\nabla \mathfrak{p}\|_q \leq C_{25} \|f\|_q \tag{3.25}$$

for a constant $C_{25} = C_{25}(n, \Omega, q, \mathcal{T}, \lambda_0) > 0$. If $(u_1, \mathfrak{p}_1) \in W^{1,2,q}_\perp(\mathbb{T} \times \Omega)^n \times L^q_\perp(\mathbb{T}; D^{1,q}(\Omega))$ is another solution to (3.24), then $u = u_1$ and $\mathfrak{p} = \mathfrak{p}_1 + \mathfrak{p}_0$ for some spatially constant function $\mathfrak{p}_0: \mathbb{T} \to \mathbb{R}$.

Proof. The result for $n = 3$ has been established in [50, Theorem 5.1]. The general case $n \geq 2$ is proved along the same lines. $\square$

Remark 3.2.2. Observe that Theorem 3.2.1 also includes the Stokes case ($\lambda = 0$). One can thus obtain well-posedness for the time-periodic Stokes problem by combining Theorem 3.2.1 with the well-known theory for the steady-state Stokes problem in an exterior domain, which can be found in [42, Chapter V] for example.

3.2.2 The Full Time-Periodic Problem

In order to treat the time-periodic Oseen problem (3.24) for general right-hand sides f, that is, without the restriction $\mathcal{P}f = 0$, we use the projection $\mathcal{P}$ to decompose problem (3.24) into two *independent* problems: The steady-state problem (3.1) and the purely periodic problem examined in the previous section. As two different well-posedness results for the steady-state problem are available (Theorem 3.1.1 and Theorem 3.1.10), we conclude two different well-posedness results for the time-periodic problem by combining each of them with Theorem 3.2.1.

We split the solution $(u, \mathfrak{p})$ into steady-state part (v, p) and purely periodic part $(w, \mathfrak{q})$ given by

$$v := \mathcal{P}u, \qquad p := \mathcal{P}\mathfrak{p}, \qquad w := \mathcal{P}_\perp u, \qquad \mathfrak{q} := \mathcal{P}_\perp \mathfrak{p}.$$

By application of $\mathcal{P}$ and $\mathcal{P}_\perp$ to (3.24), the problem is decomposed into the steady-state problem

$$\begin{cases} -\Delta v - \lambda \partial_1 v + \nabla p = \mathcal{P}f & \text{in } \Omega, \\ \operatorname{div} v = 0 & \text{in } \Omega, \\ v = 0 & \text{on } \partial\Omega, \end{cases} \qquad (3.26)$$

and the purely periodic problem

$$\begin{cases} \partial_t w - \Delta w - \lambda \partial_1 w + \nabla \mathfrak{q} = \mathcal{P}_\perp f & \text{in } \mathbb{T} \times \Omega, \\ \operatorname{div} w = 0 & \text{in } \mathbb{T} \times \Omega, \\ w = 0 & \text{on } \mathbb{T} \times \partial\Omega, \end{cases} \qquad (3.27)$$

which we treat separately.

The following theorem is a combination of Theorem 3.1.1 and Theorem 3.2.1, and it treats right-hand sides $f \in L^q(\mathbb{T} \times \Omega)$. However, the steady-state part $\mathcal{P}u$ of the velocity field u merely belongs to homogeneous Sobolev spaces as in Theorem 3.1.1.

Theorem 3.2.3. *Let $\Omega \subset \mathbb{R}^n$, $n \geq 2$ be an exterior domain of class C^2, and let $q \in (1, \frac{n+1}{2})$ and $0 < \lambda \leq \lambda_0$. Then, for every $f \in L^q(\mathbb{T} \times \Omega)^n$ there is a solution $(u, \mathfrak{p}) = (v + w, p + \mathfrak{q})$ to (3.24) with*

$$v \in X_\lambda^q(\Omega), \qquad\qquad p \in D^{1,q}(\Omega),$$
$$w \in W_\perp^{1,2,q}(\mathbb{T} \times \Omega)^n, \qquad \mathfrak{q} \in L_\perp^q(\mathbb{T}; D^{1,q}(\Omega)),$$

which satisfies

$$\|v\|_{X_\lambda^q(\Omega)} + \|\nabla p\|_q \le C_{26}|\mathcal{P}f|_q, \tag{3.28}$$

$$\|w\|_{1,2,q} + \|\nabla \mathfrak{q}\|_q \le C_{27}\|\mathcal{P}_\perp f\|_q \tag{3.29}$$

for constants $C_{26} = C_{26}(n, \Omega, q, \lambda) > 0$ and $C_{27} = C_{27}(n, \Omega, q, \mathcal{T}, \lambda_0) > 0$. However, if $q \in (1, \frac{n}{2})$, then $C_{26} = C_{26}(n, \Omega, q, \lambda_0)$, that is, C_{26} is independent of λ.

Moreover, if $(u_1, \mathfrak{p}_1)$ is another solution to (3.24) that belongs to the same function class as $(u, \mathfrak{p})$, then $u = u_1$ and $\mathfrak{p} = \mathfrak{p}_1 + \mathfrak{p}_0$ for some spatially constant function $\mathfrak{p}_0 \colon \mathbb{T} \to \mathbb{R}$.

Proof. For existence, let $f \in L^q(\mathbb{T} \times \Omega)$. Then we have $\mathcal{P}f \in L^q(\Omega)$ and $\mathcal{P}_\perp f \in L^q_\perp(\mathbb{T} \times \Omega)$. By Theorem 3.1.1 and Theorem 3.2.1 there exist solutions (v, p) and $(w, \mathfrak{q})$ to (3.26) and (3.27) that satisfy (3.28) and (3.29), respectively. Moreover, $(u, \mathfrak{p}) = (v + w, p + \mathfrak{q})$ is a solution to the time-periodic problem (3.24). To show the uniqueness statement, let $(u, \mathfrak{p})$ be the difference of two solutions for the same right-hand side f, that is, a solution to (3.24) for $f = 0$. Then the steady-state part (v, p) and the purely periodic part $(w, \mathfrak{q})$ satisfy (3.26) and (3.27) with $\mathcal{P}f = 0$ and $\mathcal{P}_\perp f = 0$, respectively. The uniqueness statement then follows from the corresponding result in Theorem 3.1.1 and Theorem 3.2.1. $\qquad\square$

Remark 3.2.4. A different combination of Theorem 3.1.1 and Theorem 3.2.1 would yield a different well-posedness result for the time-periodic problem (3.24). For example, one can extend Theorem 3.2.3 to the case

$$f \in L^{q_1}(\Omega) \otimes L^{q_2}_\perp(\mathbb{T} \times \Omega)$$

even if $q_1 \ne q_2$ by combining Theorem (3.1.1) for $q = q_1$ with Theorem (3.2.1) for $q = q_2$. One could also impose additional conditions, for example, in the purely periodic component and to consider

$$f \in L^{q_1}(\Omega) \oplus \left(L^{q_2}_\perp(\mathbb{T} \times \Omega) \cap L^{q_3}_\perp(\mathbb{T} \times \Omega)\right).$$

This can be used to derive existence of a time-periodic solution to the Navier–Stokes equations, as was shown in [50].

In the very same way, we next combine Theorem 3.1.10 and Theorem 3.2.1. Here, besides $f \in L^q(\mathbb{T} \times \Omega)$ we demand $\mathcal{P}f \in D_0^{-1,r}(\Omega)$ from the right-hand side. We then obtain a solution $(u, \mathfrak{p})$ where also the steady-state velocity field $\mathcal{P}u$ belongs to the full Sobolev space $W^{2,q}(\Omega)$.

Theorem 3.2.5. *Let* $\Omega \subset \mathbb{R}^n$, $n \geq 2$, *and let* $q \in (1, \infty)$, $r \in \left(\frac{n+1}{n}, n+1\right)$ *and* $0 < \lambda \leq \lambda_0$. *Set* $s := \frac{(n+1)r}{n+1-r}$. *Then, for every* $f \in L^q(\mathbb{T} \times \Omega)^n$ *with* $\mathcal{P}f \in D^{-1,r}(\Omega)^n$ *there is a solution* $(u, \mathfrak{p}) = (v + w, p + \mathfrak{q})$ *to* (3.24) *with*

$$v \in D^{2,q}(\Omega)^n \cap D^{1,r}(\Omega)^n \cap L^s(\Omega)^n, \qquad p \in D^{1,q}(\Omega),$$
$$w \in W_\perp^{1,2,q}(\mathbb{T} \times \Omega)^n, \qquad\qquad \mathfrak{q} \in L_\perp^q(\mathbb{T}; D^{1,q}(\Omega)),$$

which satisfies

$$|v|_{1,r} + \lambda^{\frac{1+\delta}{n+1}} \|v\|_s \leq C_{28}\lambda^{-\frac{M}{n+1}}|\mathcal{P}f|_{-1,r}, \tag{3.30}$$

$$|v|_{2,q} + \lambda\|\partial_1 v\|_q + \|\nabla p\|_q \leq C_{28}\big(\|\mathcal{P}f\|_q + \lambda^{-\frac{M}{n+1}}|\mathcal{P}f|_{-1,r}\big), \tag{3.31}$$

$$\|w\|_{1,2,q} + \|\nabla\mathfrak{q}\|_q \leq C_{28}\|\mathcal{P}_\perp f\|_q \tag{3.32}$$

for a constant $C_{28} = C_{28}(q, r, \Omega, \lambda_0, \mathcal{T}) > 0$ *and* M *and* δ *as in* (3.20). *Moreover, if* $(u_1, \mathfrak{p}_1)$ *is another solution to* (3.24) *that belongs to the same function class as* $(u, \mathfrak{p})$, *then* $u = u_1$ *and* $\mathfrak{p} = \mathfrak{p}_1 + \mathfrak{p}_0$ *for some spatially constant function* $\mathfrak{p}_0 \colon \mathbb{T} \to \mathbb{R}$.

In particular, if $s \leq q$, *then* $u \in W^{1,2,q}(\mathbb{T} \times \Omega)^n$ *and*

$$\lambda^{\frac{(1+\delta)\theta}{n+1}} \|v\|_q + \lambda^{\frac{(1+\delta)\theta}{2(n+1)}}|v|_{1,q} + |v|_{2,q} \leq C_{28}\big(\|\mathcal{P}f\|_q + \lambda^{-\frac{M}{n+1}}|\mathcal{P}f|_{-1,r}\big) \tag{3.33}$$

where $\theta \in [0, 1]$ *is given in* (3.22).

Proof. For existence, let $f \in L^q(\mathbb{T} \times \Omega)$. Then $\mathcal{P}f \in L^q(\Omega) \cap D^{-1,q}(\Omega)$ and $\mathcal{P}_\perp f \in L_\perp^q(\mathbb{T} \times \Omega)$. By Theorem 3.1.10 and Theorem 3.2.1 there exist solutions (v, p) and $(w, \mathfrak{q})$ to (3.26) and (3.27), respectively, that satisfy (3.30), (3.31) and (3.32) Moreover, $(u, \mathfrak{p}) = (v + w, p + \mathfrak{q})$ is a solution to the time-periodic problem (3.24). To show the uniqueness, let $(u, \mathfrak{p})$ be the difference of two solutions for the same right-hand side f, that is, a solution to (3.24) for $f = 0$. Then the steady-state part (v, p) and the purely periodic part $(w, \mathfrak{q})$ satisfy (3.26) and (3.27) with $\mathcal{P}f = 0$ and $\mathcal{P}_\perp f = 0$, respectively. The uniqueness statement follows from the corresponding result in Theorem 3.1.10 and Theorem 3.2.1. $\qquad\square$

Remark 3.2.6. The observation made in Remark 3.1.11 applies to Theorem 3.2.5 as well: For any $q \in \left(\frac{n+1}{n-1}, \infty\right)$ we can find a suitable r such that for any $f \in L^q(\mathbb{T} \times \Omega)^n$ with $\mathcal{P}f \in D^{-1,r}(\Omega)^n$ there exists a solution with velocity field u in the inhomogeneous Sobolev space $W^{1,2,q}(\mathbb{T} \times \Omega)^n$.

Remark 3.2.7. Here we can make the same observation as in Remark 3.2.4. One can combine Theorem 3.1.10 and Theorem 3.2.1 in a different way to obtain a well-posedness result for (3.24) that allows for

$$f \in \left(L^{q_1}(\Omega)^n \cap D^{-1,r}(\Omega)^n \right) \oplus L^{q_2}_\perp(\mathbb{T} \times \Omega)^n$$

with $q_1 \neq q_2$ and suitable r. However, this is not necessary to obtain a solution to the corresponding nonlinear problem; see Theorem 3.3.4 below.

3.3 Solutions to the Navier–Stokes Equations

Here we employ the previously derived *linear* theory for the steady-state and the purely periodic Oseen systems in order to find solutions $(u, \mathfrak{p})$ to the corresponding *nonlinear* Navier–Stokes problems. Note that, as mentioned beforehand, we give appropriate conditions for f such that the velocity field u (or its steady-state part $\mathcal{P}u$ in the time-periodic setting) belongs to the full Sobolev space $W^{2,q}(\Omega)^n$. The main idea of our approach is to reformulate the nonlinear problems as fixed-point problems for the velocity field u, which we solve by means of Banach's fixed-point theorem. Observe that our approach only yields a solution in dimension $n \geq 3$.

3.3.1 The Steady-State Navier–Stokes Problem

Here we consider the steady-state Navier–Stokes problem

$$\begin{cases} -\Delta v - \lambda \partial_1 v + \nabla p + v \cdot \nabla v = f & \text{in } \Omega, \\ \operatorname{div} v = 0 & \text{in } \Omega, \\ v = -\lambda \, \mathrm{e}_1 & \text{on } \partial\Omega, \\ \lim_{|x| \to \infty} v(x) = 0, \end{cases} \tag{3.34}$$

where Ω is an exterior domain of $\mathbb{R}^n$, $n \geq 3$, and $f: \Omega \to \mathbb{R}^n$ is a given external force, whereas $v: \Omega \to \mathbb{R}^n$ and $p: \Omega \to \mathbb{R}$ are the unknown velocity and pressure fields, respectively. We consider the case of non-vanishing velocity "at infinity", which corresponds to a non-vanishing Reynolds number $\lambda \neq 0$. Without loss of generality, we may assume $\lambda > 0$.

To show existence of a solution to (3.34) under the assumption of "small" data, we combine Theorem 3.1.10 with the contraction mapping theorem. For this purpose, we first reformulate (3.34) as a problem with homogeneous boundary conditions by means of a classical lifting procedure, which relies on the following simple lemma.

Lemma 3.3.1. *Let $R > \delta(\Omega^c)$. There exists a function $V \in C_0^\infty(\mathbb{R}^n)^n$ with* supp $V \subset B_R$, div $V \equiv 0$ *and* $V|_{\partial\Omega} = e_1$.

Proof. Consider a second radius $R_0 > 0$ with $\delta(\Omega^c) < R_0 < R$, and let $\varphi \in C_0^\infty(\mathbb{R}^n)$ with $\varphi \equiv 1$ on B_{R_0} and $\varphi \equiv 0$ on B^R. We define the function $V : \mathbb{R}^n \to \mathbb{R}^n$ by

$$V(x) := \frac{1}{2}[\Delta - \nabla \operatorname{div}](\varphi(x)x_2^2 e_1).$$

Then $V \in C_0^\infty(\mathbb{R}^n)^n$, supp $V \subset B_R$ and

$$\operatorname{div} V(x) = \frac{1}{2}[\operatorname{div}\Delta - \Delta\operatorname{div}](\varphi(x)x_2^2 e_1) = 0.$$

Moreover, for $x \in \partial\Omega$ we have $|x| < R_0$, and therefore

$$V(x) = \frac{1}{2}[\Delta - \nabla\operatorname{div}](x_2^2 e_1) = \frac{1}{2}[2 e_1 - 0] = e_1.$$

This completes the proof. $\qquad\square$

Now we set $u := v + \lambda V$, $\mathfrak{p} := p$. Then (v, p) is a solution to (3.34) if and only if $(u, \mathfrak{p})$ is a solution to

$$\begin{cases} -\Delta u - \lambda\partial_1 u + \nabla\mathfrak{p} = f + \mathcal{N}_\lambda(u) & \text{in } \Omega, \\ \operatorname{div} u = 0 & \text{in } \Omega, \\ u = 0 & \text{on } \partial\Omega, \\ \lim_{|x|\to\infty} u(x) = 0, \end{cases} \qquad (3.35)$$

where

$$\mathcal{N}_\lambda(u) = -(u - \lambda V)\cdot\nabla(u - \lambda V) - \lambda\Delta V - \lambda^2\partial_1 V. \qquad (3.36)$$

In order to show existence of a solution to the new system (3.35), we first introduce an appropriate functional framework. For $q \in (1, \infty)$ and $r \in (\frac{n+1}{n}, n+1)$ we define the function space

$$\mathcal{X}_\lambda^{q,r}(\Omega) := \{u \in D^{2,q}(\Omega)^n \cap D^{1,r}(\Omega)^n \cap L^s(\Omega)^n \mid \operatorname{div} u = 0, \, u|_{\partial\Omega} = 0\},$$

equipped with the norm

$$\|u\|_{\mathcal{X}_\lambda^{q,r}} := |u|_{2,q} + |u|_{1,r} + \lambda^{\frac{1}{n+1}}\|u\|_s, \qquad s := \frac{(n+1)r}{n+1-r}. \qquad (3.37)$$

Then $\mathcal{X}_\lambda^{q,r}(\Omega)$ is a Banach space. We introduce the solution operator

$$\mathcal{S}_\lambda : L^q(\Omega)^n \cap D_0^{-1,r}(\Omega)^n \to \mathcal{X}_\lambda^{q,r}(\Omega), \qquad f \mapsto u, \tag{3.38}$$

that maps $f \in L^q(\Omega)^n \cap D_0^{-1,r}(\Omega)^n$ to the unique velocity field $u \in \mathcal{X}_\lambda^{q,r}(\Omega)$ of a solution $(u, \mathfrak{p})$ to the steady-state Oseen problem (3.1) existing due to Theorem 3.1.10. This defines a family of continuous linear operators with

$$\|\mathcal{S}_\lambda f\|_{\mathcal{X}_\lambda^{q,r}} \leq C_{23}\big(\|f\|_q + \lambda^{-\frac{M}{n+1}}|f|_{-1,r}\big), \tag{3.39}$$

with M as in (3.20) and C_{23} independent of $\lambda \in (0, \lambda_0]$, which follows from estimates (3.18) and (3.19). Moreover, we see that $(u, \mathfrak{p})$ with $u \in \mathcal{X}_\lambda^{q,r}(\Omega)$ is a solution to (3.35) if and only if u satisfies the fixed-point equation

$$u = \mathcal{S}_\lambda\big(f + \mathcal{N}_\lambda(u)\big), \tag{3.40}$$

provided that f and $\mathcal{N}_\lambda(u)$ are elements of $L^q(\Omega)^n \cap D_0^{-1,r}(\Omega)^n$. The next lemma shows that the latter is satisfied for $u \in \mathcal{X}_\lambda^{q,r}(\Omega)$ if

$$\frac{1}{3q} \leq \min\Big\{\frac{1}{n}, \frac{1}{r} - \frac{1}{n+1}\Big\}, \tag{3.41}$$

$$\frac{1}{2r} \geq \max\Big\{\frac{1}{n+1}, \frac{1}{q} - \frac{2}{n}\Big\}. \tag{3.42}$$

Lemma 3.3.2. *Let $q, r \in (1, \infty)$ satisfy (3.41) and (3.42), and let $0 < \lambda \leq \lambda_0$. Then there exists a constant $C_{29} = C_{29}(n, \Omega, q, r, \lambda_0) > 0$ such that*

$$\|u_1 \cdot \nabla u_2\|_q \leq C_{29}\lambda^{-\frac{\theta}{n+1}}\|u_1\|_{\mathcal{X}_\lambda^{q,r}}\|u_2\|_{\mathcal{X}_\lambda^{q,r}}, \tag{3.43}$$

$$|u_1 \cdot \nabla u_2|_{-1,r} \leq C_{29}\lambda^{-\frac{\eta}{n+1}}\|u_1\|_{\mathcal{X}_\lambda^{q,r}}\|u_2\|_{\mathcal{X}_\lambda^{q,r}} \tag{3.44}$$

for all $u_1, u_2 \in \mathcal{X}_\lambda^{q,r}(\Omega)$, where $\theta, \eta \in [0, 2]$ satisfy

$$\Big(\frac{2}{s} + \frac{3}{n} - \frac{5}{2q}\Big)\theta = \frac{2}{s} + \frac{4}{n} - \frac{2}{q}, \qquad \Big(\frac{1}{s} + \frac{2}{n} - \frac{1}{q}\Big)\eta = \frac{1}{r} + \frac{4}{n} - \frac{2}{q}.$$

Proof. We choose $\theta_1 \in [0, 1]$ and $\theta_2 \in [\frac{1}{2}, 1]$ such that

$$\Big(\frac{1}{s} + \frac{2}{n} - \frac{1}{q}\Big)\theta_1 = \frac{1}{s} - \frac{1}{3q}, \qquad \Big(\frac{1}{s} + \frac{2}{n} - \frac{1}{q}\Big)\theta_2 = \frac{1}{s} + \frac{1}{n} - \frac{2}{3q},$$

which is possible due to condition (3.41). Note that this demands $\theta_1 = 1$ or $\theta_2 = 1$ if and only if $q = n/3$, in which case $2 - n/q < 0$, so that $\theta_1 = 1$

and $\theta_2 = 1$ is admissible. Therefore, the Gagliardo–Nirenberg inequality (Theorem 2.3.7) implies

$$\|u_1\|_{3q} \le c_0 |u_1|_{2,q}^{\theta_1} \|u_1\|_s^{1-\theta_1}, \qquad |u_2|_{1,3q/2} \le c_1 |u_2|_{2,q}^{\theta_2} \|u_2\|_s^{1-\theta_2}.$$

An application of Hölder's inequality thus yields

$$\|u_1 \cdot \nabla u_2\|_q \le \|u_1\|_{3q} \|\nabla u_2\|_{3q/2} \le c_2 |u_1|_{2,q}^{\theta_1} \|u_1\|_s^{1-\theta_1} |u_2|_{2,q}^{\theta_2} \|u_2\|_s^{1-\theta_2}$$
$$\le c_3 \lambda^{-\frac{\theta}{n+1}} \|u_1\|_{X_\lambda^{q,r}} \|u_2\|_{X_\lambda^{q,r}},$$

which is (3.43) with $\theta := 2 - \theta_1 - \theta_2$. Next consider $\eta_0 \in [0,1]$ such that

$$\left(\frac{1}{s} + \frac{2}{n} - \frac{1}{q} \right) \eta_0 = \frac{1}{s} - \frac{1}{2r},$$

which is possible due to (3.41) and (3.42). Moreover, we have $\eta_0 = 1$ if and only if $1/r = 2/q - 4/n$, in which case $2 - n/q < 0$, so that $\eta_0 = 1$ is admissible. Another application of the Gagliardo–Nirenberg inequality (Theorem 2.3.7) then implies

$$\|u_j\|_{2r} \le c_4 |u_j|_{2,q}^{\eta_0} \|u_j\|_s^{1-\eta_0}$$

for $j = 1, 2$. Now Hölder's inequality and the identity $u_1 \cdot \nabla u_2 = \operatorname{div}(u_1 \otimes u_2)$ lead to

$$|u_1 \cdot \nabla u_2|_{-1,r} \le c_5 \|u_1 \otimes u_2\|_r \le c_6 \|u_1\|_{2r} \|u_2\|_{2r} \le c_7 \lambda^{-\frac{\eta}{n+1}} \|u_1\|_{X_\lambda^{q,r}} \|u_2\|_{X_\lambda^{q,r}}$$

with $\eta := 2 - 2\eta_0$. $\qquad\square$

The following technical lemma enables us to employ a fixed-point argument in the end.

Lemma 3.3.3. *Let* $q, r \in (1, \infty)$ *satisfy* (3.41) *and* (3.42) *and*

$$r > \begin{cases} \frac{n}{n-1} & \text{if } n = 3, 4, \\ \frac{n+1}{n} & \text{if } n \ge 5. \end{cases} \tag{3.45}$$

Then $\max\{\theta, M + \eta\} < n + 1 - M$ *with* M *as in* (3.20) *and* θ, η *as in Lemma 3.3.2.*

Proof. Clearly, since $M \leq 2$, we have $\theta \leq 2 < n + 1 - M$ for all $n \geq 4$. For $n = 3$ we also have $\theta \leq 2 < n + 1 - M$ since $M \leq 1$ in this case by (3.45).

Now consider the term $M + \eta$, which clearly satisfies $M + \eta \leq 4$. By Lemma 3.3.2 we have $\eta = 2$ if and only if $r = (n + 1)/2$, which implies $M = 0$. Therefore, we have $M + \eta < 4 \leq n + 1 - M$ for all $n > 5$ since $M \leq 2$. Moreover, we have $M + \eta < 3 \leq n + 1 - M$ if $n \in \{3, 4\}$ since $M \leq 1$ in this case by (3.45).

In total, we obtain $\max\{\theta, M + \eta\} < n + 1 - M$. $\qquad\square$

After these technical preparations, we finally prove existence of a solution to (3.34) by resolving the fixed-point equation (3.40) by means of the contraction mapping principle.

Theorem 3.3.4. *Let $\Omega \subset \mathbb{R}^n$, $n \geq 3$, be an exterior domain of class C^2, and let $q, r \in (1, \infty)$ satisfy (3.41), (3.42) and (3.45). Then there exists $\lambda_0 > 0$ such that for all $0 < \lambda \leq \lambda_0$ there is $\varepsilon > 0$ such that for all $f \in L^q(\Omega)^n \cap D_0^{-1, r}(\Omega)^n$ satisfying $\|f\|_q + |f|_{-1, r} \leq \varepsilon$ there exists a pair (v, p) with*

$$v \in D^{2,q}(\Omega)^n \cap D^{1,r}(\Omega)^n \cap L^{\frac{(n+1)r}{n+1-r}}(\Omega)^n,$$
$$\partial_1 v \in L^q(\Omega)^n,$$
$$p \in D^{1,q}(\Omega)$$

satisfying (3.34). In particular, if $s \leq q$, then $v \in W^{2,q}(\Omega)^n$.

Remark 3.3.5. Note that the additional assumption $s \leq q$ is equivalent to $\frac{1}{q} \leq \frac{1}{r} - \frac{1}{n+1}$. In this case, (3.45) thus leads to the necessary conditions

$$q > \frac{n(n+1)}{n^2 - n - 1} \quad \text{if } n = 3, 4, \qquad q > \frac{n+1}{n-1} \quad \text{if } n \geq 5.$$

Proof. It suffices to show existence of a function $u \in \mathcal{X}_\lambda^{q,r}(\Omega)$ satisfying the fixed-point equation (3.40), that is, u is a fixed point of the mapping

$$\mathcal{F}_\lambda \colon \mathcal{X}_\lambda^{q,r}(\Omega) \to \mathcal{X}_\lambda^{q,r}(\Omega), \qquad u \mapsto \mathcal{S}_\lambda(f + \mathcal{N}_\lambda(u))$$

with $\mathcal{N}_\lambda(u)$ and $\mathcal{S}_\lambda$ given in (3.36) and (3.38), respectively. Note that $\mathcal{F}_\lambda$ is well defined since we have $\mathcal{N}_\lambda(u) \in L^q(\Omega) \cap D_0^{-1, r}(\Omega)$ for all $u \in \mathcal{X}_\lambda^{q,r}(\Omega)$ by Lemma 3.3.2. Now define the closed subset

$$A_\rho := \left\{ u \in \mathcal{X}_\lambda^{q,r}(\Omega) \mid \|u\|_{\mathcal{X}_\lambda^{q,r}} \leq \rho \right\}$$

of $\mathcal{X}_\lambda^{q,r}(\Omega)$, where the radius $\rho > 0$ will be chosen below. By Lemma 3.3.2, for $u \in A_\rho$ we have

$$\|\mathcal{N}_\lambda(u)\|_q \leq c_0 \lambda^{-\frac{\theta}{n+1}} \|u - \lambda V\|_{\mathcal{X}_\lambda^{q,r}}^2 + \|\lambda \Delta V + \lambda^2 \partial_1 V\|_q$$
$$\leq c_1 \left(\lambda^{-\frac{\theta}{n+1}} (\rho + \lambda)^2 + \lambda + \lambda^2 \right),$$
$$|\mathcal{N}_\lambda(u)|_{-1,r} \leq c_2 \lambda^{-\frac{\eta}{n+1}} \|u - \lambda V\|_{\mathcal{X}_\lambda^{q,r}}^2 + |\lambda \Delta V + \lambda^2 \partial_1 V|_{-1,r}$$
$$\leq c_3 \left(\lambda^{-\frac{\eta}{n+1}} (\rho + \lambda)^2 + \lambda + \lambda^2 \right),$$

where we used that $\|V\|_{\mathcal{X}_\lambda^{q,r}}$ is bounded uniformly for $0 < \lambda \leq \lambda_0$. From (3.39) we thus conclude

$$\|\mathcal{F}_\lambda(u)\|_{\mathcal{X}_\lambda^{q,r}} \leq c_4 \left(\|f + \mathcal{N}_\lambda(u)\|_q + \lambda^{-\frac{M}{n+1}} |f + \mathcal{N}_\lambda(u)|_{-1,r} \right)$$
$$\leq c_5 \left((1 + \lambda^{-\frac{M}{n+1}})\varepsilon + (\lambda^{-\frac{\theta}{n+1}} + \lambda^{-\frac{M+\eta}{n+1}})(\rho + \lambda)^2 + (1 + \lambda^{-\frac{M}{n+1}})(\lambda + \lambda^2) \right).$$

Similarly, for $u_1, u_2 \in A_\rho$ we obtain

$$\|\mathcal{F}_\lambda(u_1) - \mathcal{F}_\lambda(u_2)\|_{\mathcal{X}_\lambda^{q,r}}$$
$$\leq c_6 \left(\|\mathcal{N}_\lambda(u_1) - \mathcal{N}_\lambda(u_2)\|_q + \lambda^{-\frac{M}{n+1}} |\mathcal{N}_\lambda(u_1) - \mathcal{N}_\lambda(u_2)|_{-1,r} \right)$$
$$\leq c_7 \left(\lambda^{-\frac{\theta}{n+1}} + \lambda^{-\frac{M+\eta}{n+1}} \right) \left(\|u_1 + \lambda V\|_{\mathcal{X}_\lambda^{q,r}} + \|u_2 + \lambda V\|_{\mathcal{X}_\lambda^{q,r}} \right) \|u_1 - u_2\|_{\mathcal{X}_\lambda^{q,r}}$$
$$\leq c_8 \left(\lambda^{-\frac{\theta}{n+1}} + \lambda^{-\frac{M+\eta}{n+1}} \right) (\rho + \lambda) \|u_1 - u_2\|_{\mathcal{X}_\lambda^{q,r}}.$$

By Lemma 3.3.3 we have $\max\{\theta, M + \eta\} < n + 1 - M$. Therefore, there exists $\gamma \in \mathbb{R}$ with

$$1 \leq \frac{n+1}{n+1-M} < \gamma < \frac{n+1}{\max\{\theta, M + \eta\}}. \tag{3.46}$$

Now we choose $\varepsilon = \rho^\gamma = \lambda \leq \lambda_0 \leq 1$. Then the above estimates reduce to

$$\|\mathcal{F}_\lambda(u)\|_{\mathcal{X}_\lambda^{q,r}}$$
$$\leq c_9 \left(\rho^\gamma + \rho^{\gamma - \frac{\gamma M}{n+1}} + (\rho^{-\frac{\gamma \theta}{n+1}} + \rho^{-\gamma \frac{M+\eta}{n+1}})(\rho + \rho^\gamma)^2 + (1 + \rho^{-\frac{\gamma M}{n+1}})(\rho^\gamma + \rho^{2\gamma}) \right)$$
$$\leq c_{10} \left(\rho^{\gamma - 1} + \rho^{\gamma - \gamma \frac{M}{n+1} - 1} + \rho^{1 - \frac{\gamma \theta}{n+1}} + \rho^{1 - \gamma \frac{M+\eta}{n+1}} \right) \rho$$

and

$$\|\mathcal{F}_\lambda(u_1) - \mathcal{F}_\lambda(u_2)\|_{\mathcal{X}_\lambda^{q,r}} \leq c_{11} \left(\rho^{-\frac{\gamma \theta}{n+1}} + \rho^{-\gamma \frac{M+\eta}{n+1}} \right) (\rho + \rho^\gamma) \|u_1 - u_2\|_{\mathcal{X}_\lambda^{q,r}}$$
$$\leq c_{12} \left(\rho^{1 - \frac{\gamma \theta}{n+1}} + \rho^{1 - \gamma \frac{M+\eta}{n+1}} \right) \|u_1 - u_2\|_{\mathcal{X}_\lambda^{q,r}}.$$

By (3.46), we can choose $\lambda_0 > 0$ so small that

$$c_{10}\left(\rho^{\gamma-1} + \rho^{\gamma-\gamma\frac{M}{n+1}-1} + \rho^{1-\frac{\gamma\theta}{n+1}} + \rho^{1-\gamma\frac{M+\eta}{n+1}}\right) \le 1,$$

$$c_{12}\left(\rho^{1-\gamma\frac{\theta}{n+1}} + \rho^{1-\gamma\frac{M+\eta}{n+1}}\right) \le \frac{1}{2}$$

for all $\rho \le \lambda_0^{1/\gamma}$. This ensures that $\mathcal{F}_\lambda \colon A_\rho \to A_\rho$ is a contractive self-mapping for all $\lambda \in (0, \lambda_0]$. The contraction mapping principle thus yields the existence of a fixed point u of $\mathcal{F}_\lambda$, that is, of an element $u \in X_\lambda^{q,r}(\Omega)$ that satisfies (3.40). In particular, by the definition of S_λ, there exists a pressure field $\mathfrak{p}$ such that (3.35) is satisfied, and $\partial_1 u \in L^q(\Omega)$ and $\mathfrak{p} \in D^{1,q}(\Omega)$. Finally, since $V \in C_0^\infty(\mathbb{R}^n)$, the functions $v \coloneqq u - \lambda V$ and $p \coloneqq \mathfrak{p}$ form a solution to the original problem (3.35) with the asserted properties.

$\square$

3.3.2 The Time-Periodic Navier–Stokes Problem

Next we consider the time-periodic Navier–Stokes problem

$$\begin{cases} \partial_t v - \Delta v - \lambda \partial_1 v + \nabla p + v \cdot \nabla v = f & \text{in } \mathbb{T} \times \Omega, \\ \operatorname{div} v = 0 & \text{in } \mathbb{T} \times \Omega, \\ v = -\lambda e_1 & \text{on } \mathbb{T} \times \partial\Omega, \\ \lim_{|x| \to \infty} v(t,x) = 0 & \text{for } t \in \mathbb{T}. \end{cases} \tag{3.47}$$

As before, Ω is an exterior domain of $\mathbb{R}^n$, $n \ge 3$. The function $f \colon \mathbb{T} \times \Omega \to \mathbb{R}^n$ describes a given time-periodic external force, whereas $v \colon \mathbb{T} \times \Omega \to \mathbb{R}^n$ and $p \colon \mathbb{T} \times \Omega \to \mathbb{R}$ are the unknown (time-periodic) velocity and pressure fields, respectively. As before, we consider the case of non-vanishing fluid flow "at infinity", that is, without loss of generality, $\lambda > 0$.

As in the previous subsection, in order to show existence of a solution (v, p) to (3.47), we reformulate the system as a problem with homogeneous boundary condition by means of the function V introduced in Lemma 3.3.1. Set $u(t,x) \coloneqq v(t,x) + \lambda V(x)$ and $\mathfrak{p} \coloneqq p$. Then (v, p) is a $\mathcal{T}$-time-periodic solution to (3.47) if and only if $(u, \mathfrak{p})$ is a $\mathcal{T}$-time-periodic solution to

$$\begin{cases} \partial_t u - \Delta u - \lambda \partial_1 u + \nabla \mathfrak{p} = f + \mathcal{N}_\lambda(u) & \text{in } \mathbb{T} \times \Omega, \\ \operatorname{div} u = 0 & \text{in } \mathbb{T} \times \Omega, \\ u = 0 & \text{on } \mathbb{T} \times \partial\Omega, \\ \lim_{|x| \to \infty} u(t,x) = 0 & \text{for } t \in \mathbb{T}, \end{cases} \tag{3.48}$$

where $\mathcal{N}_\lambda(u)$ is defined as in (3.36). To show existence of a solution to this new system, we shall again employ a fixed-point argument within an appropriate functional framework. For $q \in (1, \infty)$ and $r \in (\frac{n+1}{n}, n+1)$ we define the function space

$$\mathcal{Z}_\lambda^{q,r}(\mathbb{T} \times \Omega) := \{u \in W^{1,2,q}(\mathbb{T} \times \Omega)^n \mid \mathrm{div}\, u = 0,\ u|_{\mathbb{T} \times \partial\Omega} = 0,\ \|u\|_{\mathcal{Z}_\lambda^{q,r}} < \infty\},$$

where

$$\|u\|_{\mathcal{Z}_\lambda^{q,r}} := \|\mathcal{P}u\|_{\mathcal{X}_\lambda^{q,r}} + \|\mathcal{P}_\perp u\|_{1,2,q} = |\mathcal{P}u|_{2,q} + |\mathcal{P}u|_{1,r} + \lambda^{\frac{1}{n+1}}\|\mathcal{P}u\|_s + \|\mathcal{P}_\perp u\|_{1,2,q},$$

with $\|\cdot\|_{\mathcal{X}_\lambda^{q,r}}$ and s as in (3.37). Then $\mathcal{Z}_\lambda^{q,r}(\mathbb{T}\times\Omega)$ is a Banach space provided $s \leq q$. We consider the solution operator

$$\mathcal{S}_\lambda \colon \left(L^q(\Omega)^n \cap D_0^{-1,r}(\Omega)^n\right) \oplus L_\perp^q(\mathbb{T} \times \Omega)^n \to \mathcal{Z}_\lambda^{q,r}(\mathbb{T} \times \Omega), \quad f \mapsto u \quad (3.49)$$

that maps a function $f \in L^q(\mathbb{T} \times \Omega)^n$ with $\mathcal{P}f \in D^{-1,r}(\Omega)^n$ to the unique velocity field u associated to a solution $(u, \mathfrak{p}) \in \mathcal{Z}_\lambda^{q,r}(\mathbb{T}\times\Omega) \times L^q(\mathbb{T}; D^{1,q}(\Omega))$ to the time-periodic Oseen problem (3.24) existing due to Theorem 3.2.5. This yields a family of continuous linear operators with

$$\|\mathcal{S}_\lambda f\|_{\mathcal{Z}_\lambda^{q,r}} \leq C_{28}\left(\|f\|_q + \lambda^{-\frac{M}{n+1}}|\mathcal{P}f|_{-1,r}\right), \quad (3.50)$$

with M as in (3.20) and C_{28} independent of $\lambda \in (0, \lambda_0]$, which follows from Theorem 3.2.5. Moreover, we see that $(u, \mathfrak{p})$ with $u \in \mathcal{Z}_\lambda^{q,r}(\mathbb{T} \times \Omega)$ is a solution to (3.48) if and only if u satisfies the fixed-point equation

$$u = \mathcal{S}_\lambda\left(f + \mathcal{N}_\lambda(u)\right), \quad (3.51)$$

provided that f and $\mathcal{N}_\lambda(u)$ are elements of $L^q(\mathbb{T} \times \Omega)$ with steady-state parts $\mathcal{P}f$ and $\mathcal{P}\mathcal{N}_\lambda(u)$ in $D_0^{-1,r}(\Omega)^n$. The following lemma shows that $\mathcal{N}_\lambda(u)$ belongs to this function class for $u \in \mathcal{Z}_\lambda^{q,r}(\mathbb{T} \times \Omega)$ provided that $q, r \in (1, \infty)$ satisfy

$$\frac{1}{q} \leq \min\left\{\frac{3}{n+2}, \frac{1}{r} - \frac{1}{n+1}\right\}, \quad (3.52)$$

$$\frac{1}{q} \geq \frac{1}{2r} \geq \max\left\{\frac{1}{n+1}, \frac{1}{q} - \frac{2}{n}\right\}. \quad (3.53)$$

Lemma 3.3.6. *Let $q, r \in (1, \infty)$ satisfy (3.52) and (3.53). Let $0 < \lambda \leq \lambda_0$ and $u_1, u_2 \in \mathcal{Z}_\lambda^{q,r}(\mathbb{T} \times \Omega)$. Set $v_j := \mathcal{P}u_j$, $w_j := \mathcal{P}_\perp u_j$ for $j = 1, 2$. Then there* *exist a constant* $C_{30} = C_{30}(n, \Omega, q, r, \lambda_0) > 0$ *and some* $\zeta \in [0, 1)$ *such that*

$$\|v_1 \cdot \nabla v_2\|_q \leq C_{30} \lambda^{-\frac{\theta}{n+1}} \|v_1\|_{\mathcal{X}_\lambda^{q,r}} \|v_2\|_{\mathcal{X}_\lambda^{q,r}}, \tag{3.54}$$

$$|v_1 \cdot \nabla v_2|_{-1,r} \leq C_{30} \lambda^{-\frac{\eta}{n+1}} \|v_1\|_{\mathcal{X}_\lambda^{q,r}} \|v_2\|_{\mathcal{X}_\lambda^{q,r}}, \tag{3.55}$$

$$\|w_1 \cdot \nabla w_2\|_q \leq C_{30} \|w_1\|_{1,2,q} \|w_2\|_{1,2,q}, \tag{3.56}$$

$$|\mathcal{P}(w_1 \cdot \nabla w_2)|_{-1,r} \leq C_{30} \|w_1\|_{1,2,q} \|w_2\|_{1,2,q}, \tag{3.57}$$

$$\|v_1 \cdot \nabla w_2\|_q \leq C_{30} \lambda^{-\frac{\zeta}{n+1}} \|v_1\|_{\mathcal{X}_\lambda^{q,r}} \|w_2\|_{1,2,q}, \tag{3.58}$$

$$\|w_1 \cdot \nabla v_2\|_q \leq C_{30} \lambda^{-\frac{\zeta}{n+1}} \|w_1\|_{1,2,q} \|v_2\|_{\mathcal{X}_\lambda^{q,r}}, \tag{3.59}$$

where $\theta, \eta \in [0, 2]$ *as in Lemma 3.3.2.*

Proof. Since (3.52) and (3.53) imply (3.41) and (3.42), the estimates (3.54) and (3.55) are direct consequences of Lemma 3.3.2.

To derive (3.56), we distinguish two cases. On the one hand, if $q > \max\{2, n/2\}$, the embeddings from Theorem 2.3.8 yield

$$\|w\|_{L^q(\mathbb{T}; L^\infty(\Omega))} + \|\nabla w\|_{L^\infty(\mathbb{T}; L^q(\Omega))} \leq c_0 \|w\|_{1,2,q}$$

for $w \in W^{1,2,q}(\mathbb{T} \times \Omega)$. Then Hölder's inequality implies

$$\|w_1 \cdot \nabla w_2\|_q \leq \|w_1\|_{L^q(\mathbb{T}; L^\infty(\Omega))} \|\nabla w_2\|_{L^\infty(\mathbb{T}; L^q(\Omega))} \leq c_1 \|w_1\|_{1,2,q} \|w_2\|_{1,2,q}.$$

On the other hand, if $q < (n+1)/2$, from $q \geq (n+2)/3$ and Theorem 2.3.8 (with $\alpha = 1/q$ and $\beta = 3 - (n+1)/q$) we conclude

$$\|w\|_{L^{2q}(\mathbb{T}; L^{\frac{nq}{n+1-2q}}(\Omega))} + \|\nabla w\|_{L^{2q}(\mathbb{T}; L^{\frac{nq}{2q-1}}(\Omega))} \leq c_2 \|w\|_{1,2,q}$$

for $w \in W^{1,2,q}(\mathbb{T} \times \Omega)$, which leads to

$$\|w_1 \cdot \nabla w_2\|_q \leq \|w_1\|_{L^{2q}(\mathbb{T}; L^{\frac{nq}{n+1-2q}}(\Omega))} \|\nabla w_2\|_{L^{2q}(\mathbb{T}; L^{\frac{nq}{2q-1}}(\Omega))} \leq c_3 \|w_1\|_{1,2,q} \|w_2\|_{1,2,q}.$$

In total, this shows (3.56).

For estimate (3.57), note that (3.52) and (3.53) imply $\frac{2(n+2)}{nq} - \frac{6}{n} \leq \frac{1}{r}$, and Theorem 2.3.8 (with $\alpha = 2 - n/q + n/2r$), leads to

$$\|w\|_{L^2(\mathbb{T}; L^{2r}(\Omega))} \leq c_4 \|w\|_{1,2,q},$$

so that Hölder's inequality yields

$$|\mathcal{P}(w_1 \cdot \nabla w_2)|_{-1,r} = |\text{div}\, \mathcal{P}(w_1 \otimes w_2)|_{-1,r} \leq c_5 \|\mathcal{P}(w_1 \otimes w_2)\|_{L^r(\Omega)}$$

$$\leq c_6 \|w_1\|_{L^2(\mathbb{T}; L^{2r}(\Omega))} \|w_2\|_{L^2(\mathbb{T}; L^{2r}(\Omega))} \leq c_7 \|w_1\|_{1,2,q} \|w_2\|_{1,2,q}.$$

For estimate (3.58) we again distinguish two cases. In the case $q \geq n$, (3.52) enables us to choose $\zeta_1 \in (0,1)$ such that

$$\left(\frac{1}{s} + \frac{2}{n} - \frac{1}{q}\right)\zeta_1 = \frac{1}{s} - \frac{1}{2q},$$

so that the Gagliardo–Nirenberg inequality (Theorem 2.3.7) implies

$$\|v_1\|_{2q} \leq c_8 |v_1|_{2,q}^{\zeta_1} \|v_1\|_s^{1-\zeta_1} \leq c_9 \lambda^{\frac{1-\zeta_1}{n+1}} \|v_1\|_{\mathcal{X}_\lambda^{q,r}}$$

Furthermore, Theorem 2.3.8 yields the estimate

$$\|\nabla w_2\|_{L^q(\mathbb{T};L^{2q}(\Omega))} \leq c_{10}\|w_2\|_{1,2,q},$$

and Hölder's inequality leads to (3.58) in the case $q \geq n$. If $s \leq q < n$, we decide $\zeta_2 \in (0,1)$ by

$$\left(\frac{1}{s} + \frac{2}{n} - \frac{1}{q}\right)\zeta_2 = \frac{1}{s} - \frac{1}{n}.$$

From the Gagliardo–Nirenberg inequality (Theorem 2.3.7) we then conclude

$$\|v_1\|_n \leq c_{11}|v_1|_{2,q}^{\zeta_2}\|v_1\|_s^{1-\zeta_2} \leq c_{12}\lambda^{\frac{1-\zeta_2}{n+1}}\|v_1\|_{\mathcal{X}_\lambda^{q,r}},$$

and Theorem 2.3.8 implies

$$\|\nabla w_2\|_{L^q(\mathbb{T};L^{\frac{nq}{n-q}}(\Omega))} \leq c_{13}\|w_2\|_{1,2,q}.$$

Hence, (3.58) also follows by Hölder's inequality in this case.

For the remaining inequality (3.59), in the case $q > \frac{n}{2}$ we choose $\zeta_3 \in (0,1)$ with

$$\left(\frac{1}{s} + \frac{2}{n} - \frac{1}{q}\right)\zeta_3 = \frac{1}{s} + \frac{1}{n} - \frac{1}{q}.$$

Note that $\zeta_3 \geq \frac{1}{2}$ since $q \geq s$. Then the Gagliardo–Nirenberg inequality (Theorem 2.3.7) and Theorem 2.3.8 lead to

$$\|\nabla v_2\|_q \leq c_{14}|v_2|_{2,q}^{\zeta_3}\|v_2\|_s^{1-\zeta_3} \leq c_{15}\lambda^{\frac{1-\zeta_3}{n+1}}\|v_2\|_{\mathcal{X}_\lambda^{q,r}},$$

$$\|w_1\|_{L^q(\mathbb{T};L^\infty(\Omega))} \leq c_{16}\|w_1\|_{1,2,q},$$

which implies (3.59) by Hölder's inequality. In the case $q < n$ we can use the Sobolev inequality (that is, Theorem 2.3.7 with $\theta = 1$), and Theorem 2.3.8 to deduce

$$\|\nabla v_2\|_{\frac{nq}{n-q}} \leq c_{17}|v_2|_{2,q} \leq c_{18}\|v_2\|_{\mathcal{X}_\lambda^{q,r}},$$

$$\|w_1\|_{L^q(\mathbb{T};L^n(\Omega))} \leq c_{19}\|w_1\|_{1,2,q},$$

and (3.59) also follows by Hölder's inequality in this case. Finally, we have shown all asserted estimates. ∎

By analogy to the previous subsection, to obtain a solution to the time-periodic Navier–Stokes problem (3.47), we impose an additional condition upon the range of r, which is justified by the following technical lemma.

Lemma 3.3.7. *Let $q, r \in (1, \infty)$ satisfy* (3.52) *and* (3.53) *and*

$$
r > \begin{cases} \frac{n}{n-1} & \text{if } n = 3, 4, \\ \frac{n+1}{n} & \text{if } n \geq 5. \end{cases} \tag{3.60}
$$

Then $\max\{0, M + \eta, \zeta\} < n + 1 - M$ *with M as in* (3.20) *and 0, η, ζ as in Lemma 3.3.6.*

Proof. Since (3.52) and (3.53) imply (3.41) and (3.42), we directly deduce $\max\{0, M + \eta\} < n + 1 - M$ from Lemma 3.3.3. Moreover, since $\zeta < 1$, we trivially have $\zeta < n - 1 \leq n + 1 - M$ for all $n \geq 3$. ∎

Finally, we establish existence of a solution to the time-periodic problem (3.47) by solving the fixed-point equation (3.51) in the function space $\mathcal{Z}_\lambda^{q,r}(\mathbb{T} \times \Omega)$.

Theorem 3.3.8. *Let $\Omega \subset \mathbb{R}^n$, $n \geq 3$, be an exterior domain of class C^2, and let $q, r \in (1, \infty)$ satisfy* (3.52), (3.53) *and* (3.60). *Then there exists $\lambda_0 > 0$ such that for all $0 < \lambda \leq \lambda_0$ there is $\varepsilon > 0$ such that for all $f \in L^q(\mathbb{T} \times \Omega)^n$ with $\mathcal{P}f \in D_0^{-1,r}(\Omega)^n$ satisfying $\|f\|_q + |\mathcal{P}f|_{-1,r} \leq \varepsilon$ there exists a solution*

$$
(v, p) \in \mathcal{Z}_\lambda^{q,r}(\mathbb{T} \times \Omega) \times L^q(\mathbb{T}; D^{1,q}(\Omega))
$$

to (3.47). *In particular, $v \in W^{1,2,q}(\mathbb{T} \times \Omega)^n$.*

Proof. The proof works similar to the proof of Theorem 3.3.4. At first, let us show existence of a function $u \in \mathcal{Z}_\lambda^{q,r}(\mathbb{T} \times \Omega)$ satisfying the fixed-point equation (3.51), that is, u is a fixed point of the mapping

$$
\mathcal{F}_\lambda \colon \mathcal{Z}_\lambda^{q,r}(\mathbb{T} \times \Omega) \to \mathcal{Z}_\lambda^{q,r}(\mathbb{T} \times \Omega), \quad u \mapsto \mathcal{S}_\lambda(f + \mathcal{N}_\lambda(u))
$$

with $\mathcal{N}_\lambda(u)$ and $\mathcal{S}_\lambda$ given in (3.36) and (3.49), respectively. Now consider the closed subset

$$
A_\rho := \left\{ u \in \mathcal{Z}_\lambda^{q,r}(\mathbb{T} \times \Omega) \mid \|u\|_{\mathcal{Z}_\lambda^{q,r}} \leq \rho \right\}
$$

of $\mathcal{Z}_\lambda^{q,r}(\mathbb{T}\times\Omega)$, where the radius $\rho > 0$ will be chosen below. For $u \in A_\rho$ set $v := \mathcal{P}u$, $w := \mathcal{P}_\perp u$. In particular, we then have

$$\mathcal{P}\mathcal{N}_\lambda(u) = -(v - \lambda V) \cdot \nabla(v - \lambda V) - \mathcal{P}(w \cdot \nabla w) - \lambda\Delta V - \lambda^2\partial_1 V,$$
$$\mathcal{P}_\perp\mathcal{N}_\lambda(u) = -(v - \lambda V) \cdot \nabla w - w \cdot \nabla(v - \lambda V) - \mathcal{P}_\perp(w \cdot \nabla w).$$

Therefore, Lemma 3.3.6 implies

$$\|\mathcal{N}_\lambda(u)\|_q \le \|\mathcal{P}\mathcal{N}_\lambda(u)\|_q + \|\mathcal{P}_\perp\mathcal{N}_\lambda(u)\|_q$$
$$\le c_0\big(\lambda^{-\frac{\theta}{n+1}}\|v - \lambda V\|_{\mathcal{X}_\lambda^{q,r}}^2 + \lambda^{-\frac{\zeta}{n+1}}\|v - \lambda V\|_{\mathcal{X}_\lambda^{q,r}}\|w\|_{1,2,q}$$
$$+ \|w\|_{1,2,q}^2 + \|\lambda\Delta V + \lambda^2\partial_1 V\|_q\big)$$
$$\le c_1\big(\lambda^{-\frac{\theta}{n+1}}(\rho + \lambda)^2 + \lambda^{-\frac{\zeta}{n+1}}(\rho + \lambda)\rho + \rho^2 + \lambda + \lambda^2\big),$$
$$|\mathcal{P}\mathcal{N}_\lambda(u)|_{-1,r} \le c_2\big(\lambda^{-\frac{\eta}{n+1}}\|u - \lambda V\|_{\mathcal{X}_\lambda^{q,r}}^2 + \|w\|_{1,2,q}^2 + |\lambda\Delta V + \lambda^2\partial_1 V|_{-1,r}\big)$$
$$\le c_3\big(\lambda^{-\frac{\eta}{n+1}}(\rho + \lambda)^2 + \rho^2 + \lambda + \lambda^2\big),$$

where we used that $\|V\|_{\mathcal{X}_\lambda^{q,r}}$ is bounded uniformly for $0 < \lambda \le \lambda_0$. From (3.50) we thus conclude

$$\|\mathcal{F}_\lambda(u)\|_{\mathcal{Z}_\lambda^{q,r}} \le c_4\big(\|f + \mathcal{N}_\lambda(u)\|_q + \lambda^{-\frac{M}{n+1}}|\mathcal{P}(f + \mathcal{N}_\lambda(u))|_{-1,r}\big)$$
$$\le c_5\big((1 + \lambda^{-\frac{M}{n+1}})\varepsilon + (\lambda^{-\frac{\theta}{n+1}} + \lambda^{-\frac{M+\eta}{n+1}})(\rho + \lambda)^2 + \lambda^{-\frac{\zeta}{n+1}}(\rho + \lambda)\rho$$
$$+ (1 + \lambda^{-\frac{M}{n+1}})(\rho^2 + \lambda + \lambda^2)\big)$$
$$\le c_6\big((1 + \lambda^{-\frac{M}{n+1}})(\varepsilon + \rho^2 + \lambda + \lambda^2) + (\lambda^{-\frac{\theta}{n+1}} + \lambda^{-\frac{M+\eta}{n+1}} + \lambda^{-\frac{\zeta}{n+1}})(\rho + \lambda)^2\big).$$

Similarly, for $u_1, u_2 \in A_\rho$ we obtain

$$\|\mathcal{F}_\lambda(u_1) - \mathcal{F}_\lambda(u_2)\|_{\mathcal{Z}_\lambda^{q,r}}$$
$$\le c_7\big(\|\mathcal{N}_\lambda(u_1) - \mathcal{N}_\lambda(u_2)\|_q + \lambda^{-\frac{M}{n+1}}|\mathcal{P}(\mathcal{N}_\lambda(u_1) - \mathcal{N}_\lambda(u_2))|_{-1,r}\big)$$
$$\le c_8\big(1 + \lambda^{-\frac{\theta}{n+1}} + \lambda^{-\frac{\zeta}{n+1}} + \lambda^{-\frac{M}{n+1}} + \lambda^{-\frac{M+\eta}{n+1}}\big)$$
$$\times (\|u_1\|_{\mathcal{Z}_\lambda^{q,r}} + \|u_2\|_{\mathcal{Z}_\lambda^{q,r}} + \|\lambda V\|_{\mathcal{X}_\lambda^{q,r}})\|u_1 - u_2\|_{\mathcal{Z}_\lambda^{q,r}}$$
$$\le c_9\big(1 + \lambda^{-\frac{\theta}{n+1}} + \lambda^{-\frac{\zeta}{n+1}} + \lambda^{-\frac{M}{n+1}} + \lambda^{-\frac{M+\eta}{n+1}}\big)(\rho + \lambda)\|u_1 - u_2\|_{\mathcal{Z}_\lambda^{q,r}}.$$

Lemma 3.3.7 implies $\max\{\theta, \zeta, M + \eta\} < n + 1 - M$, so that we can consider $\gamma \in \mathbb{R}$ with

$$1 \le \frac{n+1}{n+1-M} < \gamma < \frac{n+1}{\max\{\theta, \zeta, M + \eta\}}. \tag{3.61}$$

Now we choose $\varepsilon = \rho^\gamma = \lambda \leq \lambda_0 \leq 1$. Then the above estimates reduce to

$$\|\mathcal{F}_\lambda(u)\|_{\mathcal{Z}_\lambda^{q,r}}$$
$$\leq c_{10}\Big(\big(1 + \rho^{-\gamma\frac{M}{n+1}}\big)\big(\rho^\gamma + \rho^2 + \rho^{2\gamma}\big) + \big(\rho^{-\frac{\gamma\theta}{n+1}} + \rho^{-\gamma\frac{M+\eta}{n+1}} + \rho^{-\frac{\gamma\zeta}{n+1}}\big)\big(\rho + \rho^\gamma\big)^2\Big),$$
$$\leq c_{11}\Big(\rho^{\gamma-1} + \rho^{\gamma-\gamma\frac{M}{n+1}-1} + \rho^{1-\frac{\gamma\theta}{n+1}} + \rho^{1-\gamma\frac{M+\eta}{n+1}} + \rho^{1-\frac{\gamma\zeta}{n+1}}\Big)\rho$$

and

$$\|\mathcal{F}_\lambda(u_1) - \mathcal{F}_\lambda(u_2)\|_{\mathcal{Z}_\lambda^{q,r}}$$
$$\leq c_{12}\Big(1 + \rho^{-\frac{\gamma\theta}{n+1}} + \rho^{-\frac{\gamma\zeta}{n+1}} + \rho^{-\frac{\gamma M}{n+1}} + \rho^{-\gamma\frac{M+\eta}{n+1}}\Big)\big(\rho + \rho^\gamma\big)\|u_1 - u_2\|_{\mathcal{Z}_\lambda^{q,r}}$$
$$\leq c_{13}\Big(\rho + \rho^{1-\frac{\gamma\theta}{n+1}} + \rho^{1-\frac{\gamma\zeta}{n+1}} + \rho^{1-\frac{\gamma M}{n+1}} + \rho^{1-\gamma\frac{M+\eta}{n+1}}\Big)\|u_1 - u_2\|_{\mathcal{Z}_\lambda^{q,r}}.$$

By (3.61), we can choose $\lambda_0 > 0$ so small that

$$c_{11}\Big(\rho^{\gamma-1} + \rho^{\gamma-\gamma\frac{M}{n+1}-1} + \rho^{1-\frac{\gamma\theta}{n+1}} + \rho^{1-\gamma\frac{M+\eta}{n+1}} + \rho^{1-\frac{\gamma\zeta}{n+1}}\Big) \leq 1,$$
$$c_{13}\Big(\rho + \rho^{1-\frac{\gamma\theta}{n+1}} + \rho^{1-\frac{\gamma\zeta}{n+1}} + \rho^{1-\frac{\gamma M}{n+1}} + \rho^{1-\gamma\frac{M+\eta}{n+1}}\Big) \leq \frac{1}{2}$$

for all $\rho \leq \lambda_0^{1/\gamma}$. This ensures that $\mathcal{F}_\lambda \colon A_\rho \to A_\rho$ is a contractive self-mapping for all $\lambda \in (0, \lambda_0]$. The contraction mapping principle thus yields the existence of a fixed point u of $\mathcal{F}_\lambda$, that is, of $u \in \mathcal{Z}_\lambda^{q,r}(\Omega)$ that satisfies (3.51). Exactly as in the proof of Theorem 3.3.4, we finally see that there exists a pressure field p such that $v := u - \lambda V$ and p form a solution to the original problem (3.48) with the asserted properties. $\qquad\square$

Remark 3.3.9. Note that the conditions (3.52), (3.53) and (3.60) contain implicit assumptions on the permitted range of q aside from $\frac{n+2}{3} \leq q < n+1$. In particular, the lower bound on r from (3.60) leads to

$$\frac{1}{q} < \begin{cases} \frac{n^2-n-1}{n(n+1)} & \text{if } n = 3, 4, \\ \frac{n-1}{n+1} & \text{if } n \geq 5, \end{cases} \qquad \frac{1}{q} < \begin{cases} \frac{n+3}{2n} & \text{if } n = 3, 4, \\ \frac{(n+2)^2}{2n(n+1)} & \text{if } n \geq 5, \end{cases}$$

when combined with (3.52) and (3.53), respectively. However, by a simple calculation one verifies that this is a new condition only in the case $n = 3$. In summary, Theorem 3.3.8 provides existence of a solution (v, p) to (3.47) with $v \in W^{1,2,q}(\mathbb{T} \times \Omega)^n$ for $q \in (\frac{12}{5}, 4]$ if $n = 3$, and for $q \in [\frac{n+2}{3}, n + 1]$ if $n \geq 4$.

4 Flow Past a Rotating Body

Subject of this chapter is the investigation of the fluid flow around a rigid body $\mathcal{B}$ that moves through an infinite three-dimensional container of liquid. The velocity field that describes the rigid motion of the body shall be given by

$$V(t,x) = \xi(t) + \eta \wedge (x - x_C(t))$$

with respect to its center of mass x_C. As customary, $t \in \mathbb{R}$ and $x \in \mathbb{R}^3$ denote time and spatial variables, respectively. The quantities $\xi := \frac{\mathrm{d}}{\mathrm{d}t} x_C$ and η are the translational velocity and the angular velocity of $\mathcal{B}$ with respect to its center of mass. We assume that the translational velocity ξ is periodic of some prescribed period $\mathcal{T}$, so that we can interpret it as a function $\xi : \mathbb{T} \to \mathbb{R}^3$ for the torus group $\mathbb{T} = \mathbb{R}/\mathcal{T}\mathbb{Z}$. We consider the case where the axes of translation and rotation of the prescribed time-periodic motion of $\mathcal{B}$ are constant in time and parallel, and without loss of generality, both are directed along the x_1-axis such that

$$\xi(t) = \alpha(t)\,e_1, \qquad \eta = \omega\,e_1$$

for a function $\alpha \colon \mathbb{T} \to \mathbb{R}$ and a constant $\omega \in \mathbb{R}$. Under these conditions the time-periodic motion of an incompressible Navier–Stokes fluid around $\mathcal{B}$ that adheres to $\mathcal{B}$ at the boundary is governed by the equations

$$
\begin{cases}
\partial_t u + \omega(e_1 \wedge u - e_1 \wedge x \cdot \nabla u) - \alpha \partial_1 u + u \cdot \nabla u = f + \Delta u - \nabla \mathfrak{p} & \text{in } \mathbb{T} \times \Omega, \\
\operatorname{div} u = 0 & \text{in } \mathbb{T} \times \Omega, \\
u = \alpha\, e_1 + \omega\, e_1 \wedge x & \text{on } \mathbb{T} \times \partial\Omega, \\
\lim\limits_{|x| \to \infty} u(t, x) = 0 & \text{for } t \in \mathbb{T},
\end{cases}
$$

where $\mathbb{T} = \mathbb{R}/\mathcal{T}\mathbb{Z}$ represents the time axis and $\Omega := \mathbb{R}^3 \setminus \overline{\mathcal{B}}$ corresponds to the exterior domain filled by the fluid flow. The functions $u \colon \mathbb{T} \times \Omega \to \mathbb{R}^3$ and $\mathfrak{p} \colon \mathbb{T} \times \Omega \to \mathbb{R}$ are time-periodic velocity and pressure fields of the fluid. For the sake of generality, in the formulation of (4.1) we additionally consider a time-periodic external force $f \colon \mathbb{T} \times \Omega \to \mathbb{R}^3$ that affects the liquid. As explained in the introduction, motivated by physical observations and our mathematical approach, we only consider the configuration when the mean translational velocity is non-zero, that is,

$$
\lambda := \int_{\mathbb{T}} \alpha(t)\, \mathrm{d}t \neq 0,
$$

and when the mean rotational velocity ω coincides with the angular frequency $2\pi/\mathcal{T}$ of the time-periodic data, that is,

$$
\omega = 2\pi/\mathcal{T}.
$$

In order to capture the latter condition directly in the equations, we perform a time scale $t \to \omega t$ that yields the nonlinear system

$$
\begin{cases}
\omega(\partial_t u + e_1 \wedge u - e_1 \wedge x \cdot \nabla u) - \alpha \partial_1 u + u \cdot \nabla u = f + \Delta u - \nabla \mathfrak{p} & \text{in } \mathbb{T} \times \Omega, \\
\operatorname{div} u = 0 & \text{in } \mathbb{T} \times \Omega, \\
u = \alpha\, e_1 + \omega\, e_1 \wedge x & \text{on } \mathbb{T} \times \partial\Omega, \\
\lim\limits_{|x| \to \infty} u(t, x) = 0 & \text{for } t \in \mathbb{T}.
\end{cases}
\tag{4.1}
$$

Now all involved functions are 2π-time periodic, and $\mathbb{T}$ denotes the associated torus group $\mathbb{T} = \mathbb{R}/2\pi\mathbb{Z}$.

In order to show existence of a solution to (4.1), we study the linear time-periodic problem

$$
\begin{cases}
\omega(\partial_t u + e_1 \wedge u - e_1 \wedge x \cdot \nabla u) - \Delta u - \lambda \partial_1 u + \nabla \mathfrak{p} = f & \text{in } \mathbb{T} \times \Omega, \\
\operatorname{div} u = 0 & \text{in } \mathbb{T} \times \Omega, \\
u = 0 & \text{on } \mathbb{T} \times \partial\Omega,
\end{cases}
\tag{4.2}
$$

and establish well-posedness in a suitable functional framework. To this end, we first investigate the corresponding resolvent problem

$$
\begin{cases}
sv + \omega(e_1 \wedge v - e_1 \wedge x \cdot \nabla v) - \Delta v - \lambda \partial_1 v + \nabla p = F & \text{in } \Omega, \\
\operatorname{div} v = 0 & \text{in } \Omega, \\
v = 0 & \text{on } \partial\Omega.
\end{cases} \tag{4.3}
$$

As was found out by FARWIG and NEUSTUPA [31, 32], this resolvent problem is *not* well posed in a classical setting of $W^{2,q}$ spaces for the resolvent parameters $s = iwk \in \mathbb{C}$, $k \in \mathbb{Z}$, which are relevant for our approach. Therefore, we have to work in a different framework of homogeneous Sobolev spaces, where we establish an existence result and a suitable resolvent estimate below. Moreover, as it turns out, for each resolvent parameter $s = iwk$, $k \in \mathbb{Z}$, we obtain a *different* solution space. We combine these spaces to deduce well-posedness of the time-periodic problem (4.2) in a framework of absolutely convergent Fourier series. The main advantage of this functional framework is that it allows to directly obtain *a priori* estimates for a solution to (4.2) from the resolvent estimates for (4.3). Since a generalization of classical inequalities from the theory of Lebesgue spaces to this functional setting is straightforward, we can then show existence of a solution to the nonlinear problem (4.1).

In Section 4.1, we examine the resolvent problem (4.3). While in the whole space a well-posedness result can be obtained by a suitable change of coordinates, in an exterior domain we use cut-off techniques and a Galerkin approach. In Section 4.2, we introduce the framework of functions with absolutely convergent Fourier series, in which we then establish well-posedness for the time-periodic problem. In Section 4.3, we reformulate (4.1) as a fixed-point equation and employ the contraction mapping principle, which finally shows existence of a time-periodic solution to the nonlinear problem (4.1).

4.1 The Resolvent Problem

In order to find a solution to the nonlinear problem (4.1), in this section we investigate the resolvent problem

$$
\begin{cases}
\omega(ikv + e_1 \wedge v - e_1 \wedge x \cdot \nabla v) - \Delta v - \lambda \partial_1 v + \nabla p = F & \text{in } \Omega, \\
\operatorname{div} v = 0 & \text{in } \Omega, \\
v = 0 & \text{on } \partial\Omega
\end{cases} \tag{4.4}
$$

for $k \in \mathbb{Z}$. At first, we examine this problem together with the time-periodic problem (4.2) in the whole space $\Omega = \mathbb{R}^n$, where we employ a suitable coordinate transform, which reduces (4.2) to a time-periodic Oseen problem without rotation terms. Subsequently, we proceed with the study of the resolvent problem (4.4) in an exterior domain, and establish a uniqueness result and suitable *a priori* estimates. After showing existence of a solution in an L² framework, we combine the previous results and show well-posedness of problem (4.4).

4.1.1 Well-Posedness in the Whole Space

Before studying problems (4.2) and (4.4) in an exterior domain, we first consider the case $\Omega = \mathbb{R}^3$. This setting has the advantage that one can change coordinates back to the non-rotating inertial frame, which reduces the study of (4.2) to an investigation of the time-periodic Oseen problem without rotation terms, which has been examined in Section 3.2. For the rest of this section, we set

$$s_1 := \frac{2q}{2-q}, \qquad s_2 := \frac{4q}{4-q}, \qquad s_3 := \frac{8q}{8-q}$$

for appropriately fixed q. These numbers will occur frequently. Recall that throughout this chapter, we consider the torus group $\mathbb{T} = \mathbb{R}/2\pi\mathbb{Z}$.

The following theorem is concerned with well-posedness of the time-periodic Oseen problem in the whole space. Observe that the constants appearing in the *a priori* estimates can be chosen independently of λ and ω as long as the ratio λ^2/ω is bounded by some prescribed constant $\theta > 0$.

Theorem 4.1.1. *Let $q \in (1,2)$ and $\lambda, \omega, \theta > 0$ with $\lambda^2 \le \theta\omega$. For every $f \in L^q(\mathbb{T} \times \mathbb{R}^3)^3$ there exists a solution $(u, \mathfrak{p}) \in \mathscr{S}'(\mathbb{T} \times \mathbb{R}^3)^{3+1}$ to*

$$\begin{cases} \omega\partial_t u - \Delta u - \lambda\partial_1 u + \nabla\mathfrak{p} = f & in \ \mathbb{T} \times \mathbb{R}^3, \\ \qquad\qquad\qquad\qquad\quad \operatorname{div} u = 0 & in \ \mathbb{T} \times \mathbb{R}^3, \end{cases} \tag{4.5}$$

with $\partial_t u, \nabla^2 u, \nabla\mathfrak{p} \in L^q(\mathbb{T} \times \mathbb{R}^3)$, and constants $C_{31} = C_{31}(q) > 0$ and $C_{32} = C_{32}(q, \theta) > 0$ such that

$$\begin{aligned} \|\nabla^2\mathcal{P}u\|_q + \lambda\|\partial_1\mathcal{P}u\|_q &+ \lambda^{1/2}\|\mathcal{P}u\|_{s_1} \\ &+ \lambda^{1/4}\|\nabla\mathcal{P}u\|_{s_2} + \|\nabla\mathcal{P}\mathfrak{p}\|_q \le C_{31}\|\mathcal{P}f\|_q, \end{aligned} \tag{4.6}$$

and

$$\omega\|\partial_t\mathcal{P}_\perp u\|_q + \|\nabla^2\mathcal{P}_\perp u\|_q + \lambda\|\partial_1\mathcal{P}_\perp u\|_q + \|\nabla\mathcal{P}_\perp\mathfrak{p}\|_q \le C_{32}\|\mathcal{P}_\perp f\|_q. \tag{4.7}$$

Additionally, if $(w, \mathfrak{q}) \in \mathscr{S}'(\mathbb{T} \times \mathbb{R}^3)^{3+1}$ is another solution to (4.5), then $P_\perp u = P_\perp w$, and $Pu - Pw$ is a polynomial in each component, and $\mathfrak{p} - \mathfrak{q} = \mathfrak{p}_0$, where $\mathfrak{p}_0(t, \cdot)$ is a polynomial for each $t \in \mathbb{T}$.

Proof. We decompose (4.5) into a steady-state and a purely periodic problem by splitting $u = u_0 + u_\perp$ and $\mathfrak{p} = \mathfrak{p}_0 + \mathfrak{p}_\perp$ with

$$u_0 := Pu, \qquad \mathfrak{p}_0 := P\mathfrak{p}, \qquad u_\perp := P_\perp u, \qquad \mathfrak{p}_\perp := P_\perp \mathfrak{p}.$$

For the steady-state part $(u_0, \mathfrak{p}_0)$ we obtain the steady-state Oseen system

$$\begin{cases} -\Delta u_0 - \lambda \partial_1 u_0 + \nabla \mathfrak{p}_0 = Pf & \text{in } \mathbb{R}^3, \\ \operatorname{div} u_0 = 0 & \text{in } \mathbb{R}^3. \end{cases}$$

The existence of a time-independent solution $(u_0, \mathfrak{p}_0)$ satisfying estimate (4.6) follows from [42, Theorem VII.4.1]. The remaining purely periodic part $(u_\perp, \mathfrak{p}_\perp)$ must solve (4.5), but with purely periodic right-hand side $P_\perp f$. We define

$$U(t, x) := u_\perp(t, \omega^{-1/2} x),$$
$$\mathfrak{P}(t, x) := \omega^{-1/2} \mathfrak{p}_\perp(t, \omega^{-1/2} x),$$
$$F(t, x) := \omega^{-1} P_\perp f(t, \omega^{-1/2} x),$$

which leads to the system

$$\begin{cases} \partial_t U - \Delta U - \widetilde{\lambda} \partial_1 U + \nabla \mathfrak{P} = F & \text{in } \mathbb{T} \times \mathbb{R}^3, \\ \operatorname{div} U = 0 & \text{in } \mathbb{T} \times \mathbb{R}^3, \end{cases}$$

where $\widetilde{\lambda} = \lambda \omega^{-1/2}$. From Theorem 5.2.6 in the following chapter, we conclude the existence of a unique solution $(U, \mathfrak{P})$ that satisfies the estimate

$$\|\partial_t U\|_q + \|\nabla^2 U\|_q + \|\widetilde{\lambda} \partial_1 U\|_q + \|\nabla \mathfrak{P}\|_q \leq P(\widetilde{\lambda}) \|F\|_q,$$

where $P: \mathbb{R} \to \mathbb{R}$ is a polynomial in $\widetilde{\lambda}$ and can thus be bounded uniformly in $\widetilde{\lambda} \in (0, \sqrt{\theta}]$. Estimate (4.7) with the asserted dependency of the constant C_{32} follows after reversing the applied scaling. The remaining uniqueness statement is a direct consequence of Lemma 5.2.5, which is established in the next chapter. $\qquad \square$

Remark 4.1.2. In the setting of Theorem 4.1.1 we can write the estimate for the steady-state part $(u_0, \mathfrak{p}_0) = (Pu, P\mathfrak{p})$ and the purely periodic part

$(u_\perp, \mathfrak{p}_\perp) = (\mathcal{P}_\perp u, \mathcal{P}_\perp \mathfrak{p})$ in a more condensed way: From the embeddings established in Theorem 2.3.9 we deduce

$$\omega^{1/4}\|u_\perp\|_{L^{s_2}(\mathbb{T};L^{s_1}(\mathbb{R}^3))} + \omega^{1/8}\|\nabla u_\perp\|_{L^{s_3}(\mathbb{T};L^{s_2}(\mathbb{R}^3))}$$
$$\leq C_{33}\big(\omega\|\partial_t u_\perp\|_{L^q(\mathbb{T}\times\mathbb{R}^3)} + \|\nabla^2 u_\perp\|_{L^q(\mathbb{T}\times\mathbb{R}^3)}\big).$$

Recalling Remark 2.3.10, we see that (4.6) and (4.7) can be formulated as

$$\omega\|\partial_t u\|_q + \|\nabla^2 u\|_q + \lambda\|\partial_1 u\|_q + \lambda^{1/2}\|u\|_{L^{s_2}(\mathbb{T};L^{s_1}(\mathbb{R}^3))}$$
$$+ \lambda^{1/4}\|\nabla u\|_{L^{s_3}(\mathbb{T};L^{s_2}(\mathbb{R}^3))} + \|\nabla\mathfrak{p}\|_q \leq C_{34}\|f\|_q \tag{4.8}$$

for a constant $C_{34} = C_{34}(q,\theta)$ as long as $\lambda^2 \leq \theta\omega$.

With Theorem 4.1.1 we now solve the linear problem (4.2) for $\Omega = \mathbb{R}^3$ and $f \in L^q(\mathbb{T}\times\mathbb{R}^3)^3$ by a suitable change of coordinates.

Theorem 4.1.3. *Let* $q \in (1,2)$ *and* $\lambda, \omega, \theta > 0$ *with* $\lambda^2 \leq \theta\omega$. *For every* $f \in L^q(\mathbb{T}\times\mathbb{R}^3)^3$ *there exists a solution* $(u,\mathfrak{p}) \in \mathscr{S}'(\mathbb{T}\times\mathbb{R}^3)^{3+1}$ *to*

$$\begin{cases} \omega(\partial_t u + e_1 \wedge u - e_1 \wedge x \cdot \nabla u) - \Delta u - \lambda\partial_1 u + \nabla\mathfrak{p} = f & in\ \mathbb{T}\times\mathbb{R}^3, \\ \operatorname{div} u = 0 & in\ \mathbb{T}\times\mathbb{R}^3, \end{cases} \tag{4.9}$$

with

$$\partial_t u + e_1 \wedge u - e_1 \wedge x \cdot \nabla u,\ \nabla^2 u,\ \partial_1 u,\ \nabla\mathfrak{p} \in L^q(\mathbb{T}\times\mathbb{R}^3).$$

Moreover, there exists a constant $C_{35} = C_{35}(q,\theta) > 0$ *such that*

$$\omega\|\partial_t u + e_1 \wedge u - e_1 \wedge x \cdot \nabla u\|_{L^q(\mathbb{T}\times\mathbb{R}^3)} + \|\nabla^2 u\|_{L^q(\mathbb{T}\times\mathbb{R}^3)}$$
$$+ \lambda\|\partial_1 u\|_{L^q(\mathbb{T}\times\mathbb{R}^3)} + \lambda^{1/2}\|u\|_{L^{s_2}(\mathbb{T};L^{s_1}(\mathbb{R}^3))} + \lambda^{1/4}\|\nabla u\|_{L^{s_3}(\mathbb{T};L^{s_2}(\mathbb{R}^3))} \tag{4.10}$$
$$+ \|\nabla\mathfrak{p}\|_{L^q(\mathbb{T}\times\mathbb{R}^3)} \leq C_{35}\|f\|_{L^q(\mathbb{T}\times\mathbb{R}^3)}.$$

Additionally, if $(w,\mathfrak{q}) \in \mathscr{S}'(\mathbb{T}\times\mathbb{R}^3)^{3+1}$ *is another solution to (4.9) with* $w \in L^r(\mathbb{T}\times\mathbb{R}^3)$ *for some* $r \in [1,\infty)$, *then* $u = w$, *and* $\mathfrak{p} - \mathfrak{q} = \mathfrak{q}_0$ *for some spatially constant function* $\mathfrak{q}_0: \mathbb{T} \to \mathbb{R}$.

Proof. Let

$$Q(t) := \begin{pmatrix} 1 & 0 & 0 \\ 0 & \cos(t) & -\sin(t) \\ 0 & \sin(t) & \cos(t) \end{pmatrix}$$

be the matrix corresponding to the rotation with angular velocity e_1. Define

$$U(t,y) := Q(t)u(t, Q(t)^\top y),$$
$$\mathfrak{P}(t,y) := \mathfrak{p}(t, Q(t)^\top y),$$
$$F(t,y) := Q(t)f(t, Q(t)^\top y)$$

with the new spatial variable $y = Q(t)x$. Then u, $\mathfrak{p}$ and f satisfy (4.9) if and only if

$$\begin{cases} \omega\partial_t U - \Delta U - \lambda\partial_1 U + \nabla\mathfrak{P} = F & \text{in } \mathbb{T} \times \mathbb{R}^3, \\ \operatorname{div} U = 0 & \text{in } \mathbb{T} \times \mathbb{R}^3. \end{cases}$$

A short calculation shows

$$\partial_t U(t,y) = Q(t)(\partial_t u(t,x) + e_1 \wedge u(t,x) - e_1 \wedge x \cdot \nabla u(t,x)),$$

and the assertions of Theorem 4.1.3 are now a direct consequence of Theorem 4.1.1 and estimate (4.8). □

Remark 4.1.4. As for the corresponding steady-state problem (see [42, Theorem VIII.8.1] for example), one can extend Theorem 4.1.3 to the case of an exterior domain Ω for $f \in L^q(\mathbb{T} \times \Omega)$. However, it is not clear to the author whether or not the constant in the resulting *a priori* estimate can then be chosen independently of λ and ω. Observe that such an independence is obtained in the functional setting of Theorem 4.2.5 below, where $f \in A(\mathbb{T}; L^q(\Omega))$. Since we solve the nonlinear problem (4.1) via a fixed-point iteration that requires λ and ω to be chosen sufficiently small since they appear as data on the right-hand side, it is crucial to obtain an estimate where the constant is independent of λ and ω.

From Theorem 4.1.3 we can extract a similar result for the resolvent problem (4.4) in the whole space. For this, we identify solutions to the resolvent problem with Fourier modes of time-periodic solutions.

Theorem 4.1.5. *Let $q \in (1,2)$, $k \in \mathbb{Z}$ and λ, ω, $\theta > 0$ with $\lambda^2 \leq \theta\omega$. For every $F \in L^q(\mathbb{R}^3)^3$ there exists a solution $(v,p) \in \mathscr{S}'(\mathbb{R}^3)^{3+1}$ to*

$$\begin{cases} \omega(ikv + e_1 \wedge v - e_1 \wedge x \cdot \nabla v) - \Delta v - \lambda\partial_1 v + \nabla p = F & \text{in } \mathbb{R}^3, \\ \operatorname{div} v = 0 & \text{in } \mathbb{R}^3, \end{cases} \tag{4.11}$$

and a constant $C_{36} = C_{36}(q,\theta) > 0$ with

$$\omega\|ikv + e_1 \wedge v - e_1 \wedge x \cdot \nabla v\|_q + \|\nabla^2 v\|_q + \lambda\|\partial_1 v\|_q$$
$$+ \lambda^{1/2}\|v\|_{s_1} + \lambda^{1/4}\|\nabla v\|_{s_2} + \|\nabla p\|_q \leq C_{36}\|F\|_q. \tag{4.12}$$

Additionally, if $(w,\mathfrak{q}) \in \mathscr{S}(\mathbb{R}^3)^{3+1}$ is another solution to (4.5) with $w \in L^r(\Omega)$ for some $r \in [1,\infty)$, then $v = w$, and $p - \mathfrak{q}$ is constant.

Proof. First consider a solution (v, p) in the described function class. Then the fields

$$u(t, x) := e^{ikt} \, v(x), \qquad \mathfrak{p}(t, x) := e^{ikt} \, p(x), \qquad f(t, x) := e^{ikt} \, F(x),$$

satisfy (4.9). Therefore, uniqueness of the pair $(v, \nabla p)$ follows from the uniqueness statement in Theorem 4.1.3. To show existence, let $F \in L^q(\mathbb{R}^3)$ and define $f \in L^q(\mathbb{T} \times \mathbb{R}^3)$ as above. Theorem 4.1.3 yields the existence of a pair $(u, \mathfrak{p})$ that solves (4.9). Then the k-th Fourier coefficients $v(x) := \mathscr{F}_{\mathbb{T}}[u(\cdot, x)](k)$ and $p(x) := \mathscr{F}_{\mathbb{T}}[\mathfrak{p}(\cdot, x)](k)$ satisfy (4.11) and belong to the correct function classes. Moreover, estimate (4.12) is a direct consequence of (4.10). □

4.1.2 Uniqueness for the Resolvent Problem

Now we begin with the investigation of the resolvent problem (4.4) in an exterior domain Ω. To show a uniqueness result, we multiply $(4.4)_1$ with the considered velocity field and employ integration by parts. To justify the integration, we first have to obtain better regularity of the solution. To this end, we use a cut-off procedure to decompose the solution into one part in a bounded domain and a second part in the whole space, where the necessary regularity results are available.

Lemma 4.1.6. *Let $\lambda \geq 0$, $\omega > 0$, $k \in \mathbb{Z}$, and let (v, p) be a distributional solution to (4.4) with $F = 0$ and $\nabla^2 v, \partial_1 v, \nabla p \in L^q(\Omega)$ for some $q \in (1, \infty)$ and $v \in L^s(\Omega)$ for some $s \in (1, \infty)$. Then $v = 0$ and p is constant.*

Proof. Consider the case $\lambda > 0$ at first. Fix a radius $R > 0$ such that $\partial B_R \subset \Omega$, and define a cut-off function $\chi_0 \in C_0^\infty(\mathbb{R}^3)$ with $\chi_0(x) = 1$ for $|x| \leq 2R$ and $\chi_0(x) = 0$ for $|x| \geq 4R$. Set

$$w := \chi_0 v - \mathfrak{B}(v \cdot \nabla \chi_0), \qquad \mathfrak{q} := \chi_0 p \tag{4.13}$$

where $\mathfrak{B}$ denotes the Bogovskiĭ operator from Theorem 2.4.2. Then

$$\begin{cases} -\Delta w + \nabla \mathfrak{q} = h & \text{in } \Omega_{4R}, \\ \operatorname{div} w = 0 & \text{in } \Omega_{4R}, \\ w = 0 & \text{on } \partial \Omega_{4R}, \end{cases}$$

with

$$h := \chi_0 \big(-\omega(ikv + e_1 \wedge v - e_1 \wedge x \cdot \nabla v) - \lambda \partial_1 v \big)$$
$$- 2\nabla \chi_0 \cdot \nabla v - \Delta \chi_0 v + \nabla \chi_0 p + \Delta \mathfrak{B}(\nabla \chi_0 \cdot v).$$

From the assumptions, we obtain $v \in W^{2,q}(\Omega_{4R})$ and $p \in W^{1,q}(\Omega_{4R})$. Classical Sobolev embeddings imply $v, \nabla v, p \in L^{\frac{3}{2}q}(\Omega_{4R})$. Taking into account the mapping properties of $\mathfrak{B}$ from Theorem 2.4.2, this implies $h \in L^r(\Omega_{4R})$ for all $1 < r \leq \frac{3}{2}q$. From Theorem 2.4.5 we obtain $w \in W^{2,r}(\Omega_{4R})$ and $\nabla q \in L^r(\Omega_{4R})$. Since $v = w$ and $p = q$ on Ω_{2R}, this yields

$$(v, p) \in W^{2,r}(\Omega_{2R}) \times W^{1,r}(\Omega_{2R}) \tag{4.14}$$

for all $1 < r \leq \frac{3}{2}q$.

Next consider another cut-off function $\chi_1 \in C^\infty(\mathbb{R}^3)$ with $\chi_1(x) = 1$ for $|x| \geq 2R$ and $\chi_1(x) = 0$ for $|x| \leq R$. As above, we define

$$u := \chi_1 v - \mathfrak{B}(v \cdot \nabla \chi_1), \qquad \mathfrak{p} := \chi_1 p, \tag{4.15}$$

which satisfy the system

$$\begin{cases} \omega(iku + e_1 \wedge u - e_1 \wedge x \cdot \nabla u) - \Delta u - \lambda \partial_1 u + \nabla \mathfrak{p} = f & \text{in } \mathbb{R}^3, \\ \operatorname{div} u = 0 & \text{in } \mathbb{R}^3, \end{cases} \tag{4.16}$$

with

$$\begin{aligned} f := {}& \omega(e_1 \wedge x \cdot \nabla \chi_1) v - 2 \nabla \chi_1 \cdot \nabla v \\ & - \Delta \chi_1 v + \lambda \partial_1 \chi_1 v + \nabla \chi_1 p + \Delta \mathfrak{B}(v \cdot \nabla \chi_1) + \lambda \partial_1 \mathfrak{B}(v \cdot \nabla \chi_1) \\ & - \omega\big(ik\mathfrak{B}(v \cdot \nabla \chi_1) + e_1 \wedge \mathfrak{B}(v \cdot \nabla \chi_1) - e_1 \wedge x \cdot \nabla \mathfrak{B}(v \cdot \nabla \chi_1)\big). \end{aligned}$$

As above, we see $f \in L^r(\mathbb{R}^3)$ for all $1 < r \leq \frac{3}{2}q$. Since we also have $u \in L^s(\mathbb{R}^3)$, Theorem 4.1.5 implies

$$iku + e_1 \wedge u - e_1 \wedge x \cdot \nabla u, \nabla^2 u, \partial_1 u, \nabla \mathfrak{p} \in L^r(\mathbb{R}^3),$$
$$\nabla u \in L^{4r/(4-r)}(\mathbb{R}^3), \qquad u \in L^{2r/(2-r)}(\mathbb{R}^3)$$

if additionally $r < 2$. Due to $v = u$ and $p = \mathfrak{p}$ on B^{2R}, we have

$$\begin{aligned} & ikv + e_1 \wedge v - e_1 \wedge x \cdot \nabla v, \nabla^2 v, \partial_1 v, \nabla p \in L^r(B^{2R}), \\ & \nabla v \in L^{4r/(4-r)}(B^{2R}), \qquad v \in L^{2r/(2-r)}(B^{2R}) \end{aligned} \tag{4.17}$$

for $1 < r \leq \frac{3}{2}q$ with $r < 2$.

We combine (4.14) and (4.17) to deduce

$$\begin{aligned} & ikv + e_1 \wedge v - e_1 \wedge x \cdot \nabla v, \nabla^2 v, \partial_1 v, \nabla p \in L^r(\Omega), \\ & \nabla v \in L^{4r/(4-r)}(\Omega), \qquad v \in L^{2r/(2-r)}(\Omega) \end{aligned}$$

for $1 < r \leq \frac{3}{2}q$ with $r < 2$. After repeating the above argument a sufficient number of times, we obtain

$$\forall r \in (1,2): \ ikv + e_1 \wedge v - e_1 \wedge x \cdot \nabla v, \ \nabla^2 v, \ \partial_1 v, \ \nabla p \in L^r(\Omega) \tag{4.18}$$

and, by a combination with the Sobolev inequality,

$$\forall r \in \left(\frac{3}{2}, 6\right): \ \nabla v \in L^r(\Omega), \qquad \forall r \in (2, \infty): \ v \in L^r(\Omega). \tag{4.19}$$

In particular, using the divergence theorem and $v = 0$ on $\partial\Omega$ we obtain

$$\mathrm{Re} \int_{\Omega_R} (e_1 \wedge x \cdot \nabla v) \cdot v^* \, dx = \int_{\Omega_R} \frac{1}{2} \mathrm{div} \left[(e_1 \wedge x)|v|^2 \right] dx$$

$$= \int_{\partial\Omega_R} \frac{1}{2} (e_1 \wedge x) \cdot \mathrm{n}|v|^2 \, dS = \int_{\partial B_R} \frac{1}{2} (e_1 \wedge x) \cdot x R^{-1} |v|^2 \, dS = 0$$

for any $R > 0$ with $\partial B_R \subset \Omega$. Passing to the limit $R \to \infty$, we obtain

$$\mathrm{Re} \int_{\Omega} (e_1 \wedge x \cdot \nabla v) \cdot v^* \, dx = 0. \tag{4.20}$$

Next consider a family of cut-off functions $\chi_\rho \in C_0^\infty(\mathbb{R}^3)$ with $\chi_\rho(x) = 1$ for $|x| \leq \rho$ and $\chi_0(x) = 0$ for $|x| \geq 2\rho$ such that $0 \leq \chi_\rho \leq 1$ and $|\nabla \chi_\rho| \leq c_0/\rho$. Let $\rho > \delta(\Omega^c)$. From (4.18) and (4.19) we deduce $\partial_1 v \cdot v^* \in L^1(\Omega)$, and integration by parts and Hölder's inequality imply

$$\left| \mathrm{Re} \int_{\Omega} \chi_\rho \partial_1 v \cdot v^* \, dx \right| = \frac{1}{2} \left| \int_{\Omega} \chi_\rho \partial_1 |v|^2 \, dx \right| = \frac{1}{2} \left| \int_{\Omega} \partial_1 \chi_\rho |v|^2 \, dx \right|$$

$$\leq \frac{c_1}{\rho} \left(\int_{B_{2\rho} \setminus B_\rho} 1 \, dx \right)^{1/5} \|v\|_{5/2}^2 \leq c_2 \rho^{-2/5} \|v\|_{5/2}^2.$$

Due to (4.19), the right-hand side is finite, and by passing to the limit $\rho \to \infty$, we obtain

$$\mathrm{Re} \int_{\Omega} \partial_1 v \cdot v^* \, dx = 0. \tag{4.21}$$

By (4.18) and (4.19), we can further multiply $(4.4)_1$ by v^* and integrate over Ω. Utilizing the identities (4.20), (4.21) and $\mathrm{Re}(e_1 \wedge v \cdot v^*) = 0$ and integrating by parts, we conclude

$$0 = \mathrm{Re} \int_{\Omega} \left(\omega(ikv + e_1 \wedge v - e_1 \wedge x \cdot \nabla v) - \Delta v + \lambda \partial_1 v + \nabla p \right) \cdot v^* \, dx$$

$$= \mathrm{Re} \int_{\Omega} i\omega k |v|^2 + |\nabla v|^2 \, dx = \int_{\Omega} |\nabla v|^2 \, dx.$$

This implies $\nabla v = 0$, and the imposed boundary conditions yield $v = 0$. Finally, $(4.4)_1$ leads to $\nabla p = 0$. This completes the proof in the case $\lambda > 0$. The proof for $\lambda = 0$ can be shown in a similar way. $\qquad\square$

4.1.3 Resolvent Estimates

Next we establish an *a priori* estimate for the solution to the resolvent problem (4.4). The following lemma is the first step into this direction. Observe that the right-hand side of estimate (4.23) below still contains lower order terms. For its derivation, we use a cut-off argument as in the proof of Lemma 4.1.6. Note that, as explained above, we have to keep record on the influence of the parameters λ and ω.

Lemma 4.1.7. *Let $q \in (1,2)$, $k \in \mathbb{Z}$ and $\lambda, \omega, \theta > 0$ with $\lambda^2 \le \theta\omega$. Moreover, let $F \in L^q(\Omega)^3$ and $R > \delta(\Omega^c)$. Let $(v,p) \in L^1_{\mathrm{loc}}(\Omega)^{3+1}$ with*

$$ikv + e_1 \wedge v - e_1 \wedge x \cdot \nabla v, \ \nabla^2 v, \ \partial_1 v, \ \nabla p \in L^q(\Omega), \tag{4.22}$$
$$v \in L^{s_1}(\Omega), \qquad \nabla v \in L^{s_2}(\Omega)$$

be a solution to (4.4). Then there is a constant $C_{37} = C_{37}(\Omega, q, \theta, R) > 0$ such that

$$\omega\|ikv + e_1 \wedge v - e_1 \wedge x \cdot \nabla v\|_q + \|\nabla^2 v\|_q$$
$$+ \lambda\|\partial_1 v\|_q + \lambda^{1/2}\|v\|_{s_1} + \lambda^{1/4}\|\nabla v\|_{s_2} + \|\nabla p\|_q \tag{4.23}$$
$$\le C_{37}\big(\|F\|_q + (1 + \lambda + \omega)\|v\|_{1,q;\Omega_{4R}} + \omega|k|\,\|v\|_{-1,q;\Omega_{4R}} + \|p\|_{q;\Omega_{4R}}\big).$$

Proof. Let χ_0, χ_1 be the cut-off functions from the proof of Lemma 4.1.6. Define $w := \chi_0 v$ and $\mathfrak{q} := \chi_0 p$. Then $w \in W^{2,q}(\Omega_{4R})$, $\mathfrak{q} \in W^{1,q}(\Omega_{4R})$ and

$$\begin{cases} ik\omega\, w - \Delta w + \nabla\mathfrak{q} = h & \text{in } \Omega_{4R}, \\ \operatorname{div} w = g & \text{in } \Omega_{4R}, \\ w = 0 & \text{on } \partial\Omega_{4R} \end{cases}$$

with

$$h := \big(F - \omega(e_1 \wedge v - e_1 \wedge x \cdot \nabla v) - \lambda\partial_1 v\big)\chi_0 - 2\nabla\chi_0 \cdot \nabla v - \Delta\chi_0 v + \nabla\chi_0 p,$$
$$g := v \cdot \nabla\chi.$$

Therefore, $(w, \mathfrak{q})$ satisfies a Stokes resolvent problem in the bounded domain Ω_{4R}, and Theorem 2.4.5 implies

$$\|w\|_{2,q;\Omega_{4R}} + \|\nabla\mathfrak{q}\|_{q;\Omega_{4R}} \le c_0\big(\|h\|_{q;\Omega_{4R}} + \|\nabla g\|_{q;\Omega_{4R}} + \omega|k|\,g\big|^*_{-1,q;\Omega_{4R}}\big),$$

where $|\cdot|^*_{-1,q;\Omega_{4R}}$ is defined in (2.11). By Hölder's inequality and Proposition 2.4.3, we obtain

$$\|h\|_{q;\Omega_{4R}} + \|\nabla g\|_{q;\Omega_{4R}} \le c_1\big(\|F\|_q + (1+\lambda+\omega)\|v\|_{1,q;\Omega_{4R}} + \|p\|_{q;\Omega_{4R}}\big),$$

$$|g|^*_{-1,q;\Omega_{4R}} \le c_2\|v\|_{-1,q;\Omega_{4R}}.$$

Since $v = w$ and $p = \mathfrak{q}$ on Ω_{2R}, we thus conclude

$$\begin{aligned}
\|v\|_{2,q;\Omega_{2R}} &+ \|\nabla p\|_{q;\Omega_{2R}} \\
&\le c_3\big(\|F\|_q + (1+\lambda+\omega)\|v\|_{1,q;\Omega_{4R}} + \|p\|_{q;\Omega_{4R}} + \omega|k|\,\|v\|_{-1,q;\Omega_{4R}}\big).
\end{aligned} \tag{4.24}$$

Next define $(u, \mathfrak{p})$ as in (4.15), which satisfies the system

$$\begin{cases}
\omega(iku + \mathrm{e}_1 \wedge u - \mathrm{e}_1 \wedge x \cdot \nabla u) - \Delta u - \lambda\partial_1 u + \nabla\mathfrak{p} = f & \text{in } \mathbb{R}^3, \\
\hfill \operatorname{div} u = 0 & \text{in } \mathbb{R}^3,
\end{cases}$$

with

$$\begin{aligned}
f := {}& \chi_1 F - \omega(\mathrm{e}_1 \wedge x \cdot \nabla\chi_1)v - 2\nabla\chi_1 \cdot \nabla u - \Delta\chi_1 v + \lambda\partial_1\chi_1 v \\
&+ \nabla\chi_1 p - \Delta\mathfrak{B}(v \cdot \nabla\chi_1) + \lambda\partial_1\mathfrak{B}(v \cdot \nabla\chi_1) \\
&+ \omega\big(ik\mathfrak{B}(v \cdot \nabla\chi_1) + \mathrm{e}_1 \wedge \mathfrak{B}(v \cdot \nabla\chi_1) - \mathrm{e}_1 \wedge x \cdot \nabla\mathfrak{B}(v \cdot \nabla\chi_1)\big),
\end{aligned}$$

where $\mathfrak{B}$ is the Bogovskiĭ operator; see Theorem 2.4.2. Theorem 4.1.5 and the mapping properties of the Bogovskiĭ operator from Theorem 2.4.2 and Corollary 2.4.4 lead to the estimate

$$\begin{aligned}
\omega\|iku &+ \mathrm{e}_1 \wedge u - \mathrm{e}_1 \wedge x \cdot \nabla u\|_q + \|\nabla^2 u\|_q \\
&+ \lambda\|\partial_1 u\|_q + \lambda^{1/4}\|\nabla u\|_{s_2} + \lambda^{1/2}\|u\|_{s_1} + \|\nabla\mathfrak{p}\|_q \\
&\le c_4\big(\|F\|_q + (1+\lambda+\omega)\|v\|_{1,q;\Omega_{2R}} + \|p\|_{q;\Omega_{2R}} + \omega|k|\,\|v\|_{-1,q;\Omega_{2R}}\big).
\end{aligned}$$

Due to $v = u$ and $p = \mathfrak{p}$ on Ω^{2R}, this implies

$$\begin{aligned}
\omega\|ikv &+ \mathrm{e}_1 \wedge v - \mathrm{e}_1 \wedge x \cdot \nabla v\|_{q;\Omega^{2R}} + \|\nabla^2 v\|_{q;\Omega^{2R}} \\
&+ \lambda\|\partial_1 v\|_{q;\Omega^{2R}} + \lambda^{1/4}\|\nabla v\|_{s_2;\Omega^{2R}} + \lambda^{1/2}\|v\|_{s_1;\Omega^{2R}} + \|\nabla p\|_{q;\Omega^{2R}} \\
&\le c_5\big(\|F\|_q + (1+\lambda+\omega)\|v\|_{1,q;\Omega_{2R}} + \|p\|_{q;\Omega_{2R}} + \omega|k|\,\|v\|_{-1,q;\Omega_{2R}}\big).
\end{aligned}$$

Combining this estimate with (4.24), we conclude (4.23). $\qquad\square$

In the next step we improve estimate (4.23) by showing that the lower-order terms on the right-hand side can be omitted. This leads to the desired *a priori* estimate, where only the data F appear on the right-hand side. For the proof we use a classical contradiction argument. Note that, because we need a constant that is independent of the parameters k, λ and ω, this argument is more involved than in the classical case.

Lemma 4.1.8. *Let $q \in (1,2)$, $k \in \mathbb{Z}$ and $\lambda, \omega > 0$, and let $F \in L^q(\Omega)^3$. Let $(v,p) \in L^1_{\mathrm{loc}}(\Omega)^{3+1}$ be a solution to (4.4) in the class (4.22). Then the estimate*

$$
\begin{aligned}
\omega \| ikv + e_1 \wedge v - e_1 \wedge x \cdot \nabla v \|_q &+ \| \nabla^2 v \|_q + \lambda \| \partial_1 v \|_q \\
&+ \lambda^{1/2} \| v \|_{s_1} + \lambda^{1/4} \| \nabla v \|_{s_2} + \| \nabla p \|_q \leq C_{38} \| F \|_q
\end{aligned}
\tag{4.25}
$$

holds for a constant $C_{38} = C_{38}(\Omega, q, \lambda, \omega) > 0$. If $q \in (1, \frac{3}{2})$ and $\lambda^2 \leq \theta \omega \leq B$ then this constant can be chosen independently of λ and ω such that $C_{38} = C_{38}(\Omega, q, \theta, B)$.

Proof. We prove the lemma by a contradiction argument. At first, consider the case $q \in (1, \frac{3}{2})$ and assume that (4.25) is not valid for a constant $C_{38} = C_{38}(\Omega, q, \theta, B)$. Then there exist sequences of numbers $(\lambda_j) \subset (0, \sqrt{B}]$, $(\omega_j) \subset (0, B/\theta]$ with $\lambda_j^2 \leq \theta \omega_j$, and $(k_j) \subset \mathbb{Z}$, and of functions (v_j), (p_j), (F_j) that satisfy

$$
\lim_{j \to \infty} \| F_j \|_q \to 0,
\tag{4.26}
$$

$$
\begin{aligned}
\omega_j \| ik_j v_j + e_1 \wedge v_j - e_1 \wedge x \cdot \nabla v_j \|_q &+ \| \nabla^2 v_j \|_q \\
+ \lambda_j \| \partial_1 v_j \|_q + \lambda_j^{1/2} \| v_j \|_{s_1} &+ \lambda_j^{1/4} \| \nabla v_j \|_{s_2} + \| \nabla p_j \|_q = 1,
\end{aligned}
\tag{4.27}
$$

and

$$
\begin{cases}
\omega_j (ik_j v_j + e_1 \wedge v_j - e_1 \wedge x \cdot \nabla v_j) - \Delta v_j - \lambda_j \partial_1 v_j + \nabla p_j = F_j & \text{in } \Omega, \\
\operatorname{div} v_j = 0 & \text{in } \Omega, \\
v_j = 0 & \text{on } \partial\Omega,
\end{cases}
\tag{4.28}
$$

for all $j \in \mathbb{N}$. Furthermore, without loss of generality, we may assume $\int_{\Omega_R} p_j \, dx = 0$ for a radius $R > \delta(\Omega^c)$. Then, (λ_j), (ω_j) and (k_j) contain convergent subsequences with limits $\lambda \in [0, \sqrt{B}]$, $\omega \in [0, B/\theta]$ and $k \in \mathbb{Z} \cup \{\pm\infty\}$, respectively, such that $\lambda^2 \leq \theta\omega$. For simplicity, we identify selected subsequences with the actual sequences for the rest of the proof.

For the moment, fix a radius $\rho > R$. By equation (4.27), the sequence (v_j) is bounded in $D^{2,q}(\Omega_\rho)$. Due to $(4.28)_3$ and the generalized Poincaré inequality from Proposition 2.3.3, the sequence (v_j) is also bounded in $W^{2,q}(\Omega_\rho)$. Moreover, this fact and the estimate

$$
\| i\omega_j k_j v_j \|_{q;\Omega_\rho} \leq \omega_j \| ik_j v_j + e_1 \wedge v_j - e_1 \wedge x \cdot \nabla v_j \|_{q;\Omega_\rho} + c_0 \omega_j (1 + \rho) \| v_j \|_{1,q;\Omega_\rho}
$$

together with (4.27) show that the sequence $(i\omega_j k_j v_j)$ is bounded in the space $L^q(\Omega_\rho)$. Furthermore, by (4.27) and Poincaré's inequality, the sequence (p_j) is bounded in $W^{1,q}(\Omega_\rho)$. In summary, we conclude that

$U_j := (i\omega_j k_j v_j, v_j, p_j)$ is bounded in $L^q(\Omega_\rho) \times W^{2,q}(\Omega_\rho) \times W^{1,q}(\Omega_\rho)$ for any $\rho > R$. Hence, by a Cantor diagonalization argument, there exists $U := (w, v, p)$ such that a subsequence of (U_j) converges weakly in $L^q(\Omega_\rho) \times W^{2,q}(\Omega_\rho) \times W^{1,q}(\Omega_\rho)$ to U for each $\rho > R$. Consequently, passing to the limit $j \to \infty$ in (4.28) and using (4.26), we obtain

$$\begin{cases} w + \omega(e_1 \wedge v - e_1 \wedge x \cdot \nabla v) - \Delta v - \lambda \partial_1 v + \nabla p = 0 & \text{in } \Omega, \\ \operatorname{div} v = 0 & \text{in } \Omega, \\ v = 0 & \text{on } \partial\Omega. \end{cases} \tag{4.29}$$

Moreover, by the compact embeddings

$$W^{2,q}(\Omega_{4R}) \hookrightarrow W^{1,q}(\Omega_{4R}) \hookrightarrow L^q(\Omega_{4R}) \hookrightarrow W^{-1,q}(\Omega_{4R}),$$

we deduce that U is the strong limit of (U_j) in the topology of the space $W^{-1,q}(\Omega_{4R}) \times W^{1,q}(\Omega_{4R}) \times L^q(\Omega_{4R})$. By Lemma 4.1.7, we conclude

$$\omega_j \|ik_j v_j + e_1 \wedge v_j - e_1 \wedge x \cdot \nabla v_j\|_q + \|\nabla^2 v_j\|_q$$
$$+ \lambda_j \|\partial_1 v_j\|_q + \lambda_j^{1/2} \|v_j\|_{s_1} + \lambda_j^{1/4} \|\nabla v_j\|_{s_2} + \|\nabla p_j\|_q$$
$$\leq C_{37} \big(\|F_j\|_q + (1 + \lambda_j + \omega_j) \|v_j\|_{1,q;\Omega_{4R}} + \omega |k_j| \, \|v_j\|_{-1,q;\Omega_{4R}} + \|p_j\|_{q;\Omega_{4R}} \big).$$

Passing to the limit $j \to \infty$ in this estimate, we conclude in virtue of (4.26) and (4.27) that

$$1 \leq C_{37} \big((1 + \lambda + \omega) \|v\|_{1,q;\Omega_{4R}} + \|w\|_{-1,q;\Omega_{4R}} + \|p\|_{q;\Omega_{4R}} \big). \tag{4.30}$$

Moreover, this implies

$$\|w + \omega(e_1 \wedge v - e_1 \wedge x \cdot \nabla v)\|_q + \|\nabla^2 v\|_q + \lambda \|\partial_1 v\|_q$$
$$+ \lambda^{1/2} \|v\|_{s_1} + \lambda^{1/4} \|\nabla v\|_{s_2} + \|\nabla p\|_q < \infty. \tag{4.31}$$

Now we distinguish between several cases:

i. If $\omega_j k_j \to s \in \mathbb{R}$ and $\omega = 0$, then $\lambda = 0$ and $w = isv$, so that (4.29) reduces to a Stokes resolvent problem. If $s \neq 0$, we also have $v \in L^q(\Omega)$ and we conclude $v = \nabla p = 0$ from a well-known uniqueness result; see for example [33]. If $s = 0$, we utilize that $q < \frac{3}{2}$ and $v_j \in L^{s_1}(\Omega)$, $\nabla v_j \in L^{s_2}(\Omega)$, so that Sobolev's inequality implies

$$\|v_j\|_{3q/(3-2q)} \leq c_1 \|\nabla v_j\|_{3q/(3-q)} \leq c_2 \|\nabla^2 v_j\|_q,$$

and thus $v \in L^{3q/(3-2q)}(\Omega)$. Now $v = \nabla p = 0$ follows from Lemma 3.1.3.

ii. If $\omega_j k_j \to s \in \mathbb{R}$ and $\omega \neq 0$ but $\lambda = 0$, then $k_j \to k \in \mathbb{Z}$ and $w = i\omega k v$, so that (4.29) reduces to the resolvent problem (4.4) with $\lambda = 0$. As above, we deduce $v \in L^{3q/(3-2q)}(\Omega)$, and from Lemma 4.1.6 we conclude $v = \nabla p = 0$.

iii. If $\omega_j k_j \to s \in \mathbb{R}$ and $\omega \neq 0$ and $\lambda \neq 0$, then $k_j \to k \in \mathbb{Z}$ and $w = i\omega k v$, so that (v, p) satisfies the resolvent problem (4.4). Since $\lambda \neq 0$, from (4.31) we obtain $v \in L^{s_1}(\Omega)$. Lemma 4.1.6 thus implies $v = \nabla p = 0$.

iv. If $\omega_j |k_j| \to \infty$, we recall (4.27) and estimate

$$\omega_j |k_j| \|v_j\|_{q;\Omega_\rho} \leq \omega_j \|ik_j v_j + e_1 \wedge v_j - e_1 \wedge x \cdot \nabla v_j\|_{q;\Omega_\rho} + c_3 \|v_j\|_{1,q;\Omega_\rho} \leq c_4,$$

for any $\rho > R$, where the constants c_3 and c_4 depend on ρ but are independent of v_j. Passing to the limit $j \to \infty$, we thus obtain $v = 0$ on Ω_ρ for each $\rho > R$. Therefore, $v = 0$ on Ω, and (4.29)$_1$ reduces to $w + \nabla p = 0$. Clearly, we also have $\operatorname{div} w = 0$ and $w|_{\partial\Omega} = 0$, so that $w + \nabla p = 0$ corresponds to the Helmholtz decomposition of 0 in $L^q(\Omega)$. Since this decomposition is unique, we conclude $w = \nabla p = 0$.

Consequently, all four cases lead to $w = v = \nabla p = 0$, which contradicts (4.30). This completes the proof in the case $1 < q < \frac{3}{2}$.

In the more general case $q \in (1, 2)$, where we do not assert the constant C_{38} to be independent of λ and ω, these parameters remain fixed in the contradiction argument. Consequently, only the last two cases above have to be considered. Since the conclusion in both of these cases is valid for all $q \in (1, 2)$, we conclude the lemma. $\qquad\square$

4.1.4 Existence of a Solution

After having established a uniqueness statement and suitable estimates, it remains to show existence of a solution to the resolvent problem (4.4). To this end, we employ a Galerkin approach combined with an "invading domains" technique to obtain a solution in an L^2 framework.

More precisely, we proceed as follows. We intersect the exterior domain Ω with balls of radius B_R. On the resulting bounded domain Ω_R, we can select eigenfunctions to the Stokes operator, which we take as a basis for a Galerkin approximation and lead to a strong solution in Ω_R. Subsequently, we pass to the limit $R \to \infty$, which finally yields a solution in the exterior domain Ω. Observe that in order for this approach to work, we need to derive suitable *a priori* estimates that are valid independently of R.

However, in contrast to Lemma 4.1.8, here it is not necessary to examine the dependencies on k, λ and ω.

During our approach we make use of the following identity.

Lemma 4.1.9. *Let $\Omega \subset \mathbb{R}^n$ be an exterior domain of class C^2, and let $R > \delta(\Omega^c)$. Let $u \in L^2_\sigma(\Omega_R) \cap W^{1,2}_0(\Omega_R) \cap W^{2,2}(\Omega_R)$ with complex conjugate u^*. Then $e_1 \wedge u - e_1 \wedge x \cdot \nabla u \in L^2_\sigma(\Omega_R)$ and*

$$\int_{\Omega_R} (e_1 \wedge u - e_1 \wedge x \cdot \nabla u) \cdot \mathcal{P}_H \Delta u^* \, dx$$

$$= \int_{\partial\Omega} \frac{1}{2} |\nabla u|^2 (e_1 \wedge x) \cdot n - n \cdot \nabla u^* \cdot (e_1 \wedge x \cdot \nabla u) \, dS - \int_{\Omega_R} \nabla(e_1 \wedge u) : \nabla u^* \, dx.$$

Proof. This was proved in [52, Lemma 3] for real-valued functions u and simply carries over to complex-valued functions by a decomposition into real and imaginary parts. ☐

Next we introduce a basis of functions suitable for our Galerkin method.

Lemma 4.1.10. *Let D be a bounded domain of class C^2. There is a sequence of (real-valued) eigenfunctions $(\psi_j)_{j \in \mathbb{N}}$ of the Stokes operator A, defined in (2.16), and a sequence $(\mu_j)_{j \in \mathbb{N}} \subset (0, \infty)$ of eigenvalues such that $(\psi_j)_{j \in \mathbb{N}}$ constitutes a basis of both $\mathrm{dom}(A)$ and $L^2_\sigma(D)$ and*

$$\mu_j \int_D \psi_j \cdot \psi_\ell \, dx = \delta_{j\ell}. \tag{4.32}$$

Proof. Since A is a positive self-adjoint and invertible operator, the existence of positive eigenvalues and corresponding eigenfunctions that constitute a basis of $\mathrm{dom}(A)$ and an orthonormal basis of $L^2(D)$ follows from classical spectral theory; see [100, Theorem VI.5.1] for example. Finally, by a renormalization we conclude (4.32). ☐

Remark 4.1.11. Observe that (4.32) means that $(\psi_j)_{j \in \mathbb{N}}$ is an orthonormal set in $W^{1,2}_0(D)^n$ equipped with the homogeneous norm $|\cdot|_{1,2}$ since

$$\mu_j \int_D \psi_j \cdot \psi_\ell \, dx = - \int_D \mathcal{P}_H \Delta \psi_j \cdot \psi_\ell \, dx = \int_D \nabla \psi_j : \nabla \psi_\ell \, dx.$$

Making use of this basis, we next show existence of a solution to system (4.4) by means of a Galerkin method. Note that a basis of eigenfunctions as in Lemma 4.1.10 does not exist in an exterior domain. Therefore, we consider (4.4) in a suitable bounded domain at first.

Lemma 4.1.12. *Let $\Omega \subset \mathbb{R}^3$ be an exterior domain of class C^3. Let $\lambda, \omega > 0$, $k \in \mathbb{Z}$, and let $F \in L^2(\Omega)^3 \cap L^{6/5}(\Omega)^3$. For each R, $R_0 > 0$ with $R > R_0 > \delta(\Omega^c)$, there exists a solution $(v^R, p^R) = (v, p)$ to*

$$
\begin{cases}
\omega(ikv + e_1 \wedge v - e_1 \wedge x \cdot \nabla v) - \Delta v - \lambda \partial_1 v + \nabla p = F & in\ \Omega_R, \\
\operatorname{div} v = 0 & in\ \Omega_R, \\
v = 0 & on\ \partial\Omega_R
\end{cases}
\tag{4.33}
$$

satisfying

$$
\begin{aligned}
\|v\|_{6;\Omega_R} + \|\nabla v\|_{2;\Omega_R} + \|\nabla^2 v\|_{2;\Omega_R} + \omega \|ikv + e_1 \wedge v - e_1 \wedge x \cdot \nabla v\|_{2;\Omega_R} \\
\leq C_{39} \big(\|F\|_{6/5;\Omega_R} + \|F\|_{2;\Omega_R} \big)
\end{aligned}
\tag{4.34}
$$

for a constant $C_{39} = C_{39}(n, \Omega, k, \lambda, \omega, R_0) > 0$ that is independent of R.

Proof. Let $(\psi_j)_{j\in\mathbb{N}}$ and $(\mu_j)_{j\in\mathbb{N}} \subset (0, \infty)$ be as in Lemma 4.1.10 for $D = \Omega_R$, and consider the set $X_n := \operatorname{span}_{\mathbb{C}}\{\psi_j \mid j = 1, \dots, n\}$. At first, we construct a function $u = u_n \in X_n$ satisfying

$$
\int_{\Omega_R} \big[\omega(iku + e_1 \wedge u - e_1 \wedge x \cdot \nabla u) - \Delta u - \lambda \partial_1 u \big] \cdot \psi_j \, \mathrm{d}x = \int_{\Omega_R} F \cdot \psi_j \, \mathrm{d}x
\tag{4.35}
$$

for all $j \in \{1, \dots, n\}$. Note that these are linear equations in the finite-dimensional vector space X_n. Since $u \in X_n$, there exist $\xi := (\xi_1, \dots, \xi_n) \in \mathbb{C}$ such that

$$
u = \sum_{\ell=1}^{n} \xi_\ell \psi_\ell.
$$

Observe that the orthogonality relation (4.32) and $\psi_j \in L^2_\sigma(\Omega_R)$ yield

$$
\int_{\Omega_R} -\Delta \psi_\ell \cdot \psi_j \, \mathrm{d}x = \int_{\Omega_R} -\mathcal{P}_{\mathrm{H}}\Delta \psi_\ell \cdot \psi_j \, \mathrm{d}x = \int_{\Omega_R} \mu_\ell \psi_\ell \cdot \psi_j \, \mathrm{d}x = \delta_{\ell j}.
$$

Therefore, with the above representation of u, the system of equations (4.35) reduces to the algebraic equation

$$
(I + iM)\xi = b
\tag{4.36}
$$

with $M = (M_{\ell j}) \in \mathbb{C}^{n \times n}$ and $b = (b_j) \in \mathbb{C}^n$ defined by

$$
M_{\ell j} := \int_{\Omega_R} \big(\omega k \psi_\ell - i\omega(e_1 \wedge \psi_\ell - e_1 \wedge x \cdot \nabla \psi_\ell) + i\lambda \partial_1 \psi_\ell \big) \cdot \psi_j \, \mathrm{d}x,
$$

$$
b_j := \int_{\Omega_R} F \cdot \psi_j \, \mathrm{d}x.
$$

Employing the vector identities

$$e_1 \wedge \psi_\ell \cdot \psi_j = -e_1 \wedge \psi_j \cdot \psi_\ell, \qquad e_1 \wedge x \cdot \nabla \psi_\ell = \operatorname{div}\left[\psi_\ell \otimes (e_1 \wedge x)\right]$$

and integration by parts, we see that M is a self-adjoint matrix and thus possesses only real eigenvalues. Since equation (4.36) is equivalent to $(-iI + M)\xi = -ib$, we thus obtain a unique solution $\xi \in \mathbb{R}^n$, which leads to existence of a unique solution $u = u_n \in X_n$ to (4.35).

Next we derive suitable estimates for $u = u_n$. Multiplication of both sides of (4.35) by the complex conjugate coefficient ξ_j^* and summation over $j = 1, \ldots, n$ yields

$$\|\nabla u\|_2^2 + \int_{\Omega_R} \left(\omega(iku + e_1 \wedge u - e_1 \wedge x \cdot \nabla u) - \lambda \partial_1 u\right) \cdot u^* \, dx = \int_{\Omega_R} F \cdot u^* \, dx.$$

Employing the above vector identities again, we see that the integral term on the left-hand side is purely imaginary. Taking the real part of this equation thus leads to the estimate

$$\|\nabla u\|_2^2 \leq \|F\|_{6/5}\|u\|_6.$$

Recalling the Sobolev inequality $\|u\|_6 \leq c_0\|\nabla u\|_2$, we obtain the estimate $\|\nabla u\|_2 \leq c_0\|F\|_{6/5}$ and conclude

$$\|u\|_6 + \|\nabla u\|_2 \leq c_1\|F\|_{6/5}, \tag{4.37}$$

where c_1 is independent of R. When we multiply both sides of (4.35) by $\mu_j \xi_j^*$ and sum over $j = 1, \ldots, n$, we obtain

$$\|\mathcal{P}_\mathrm{H}\Delta u\|_2^2 = \int_{\Omega_R} \left[F - \omega(iku + e_1 \wedge u - e_1 \wedge x \cdot \nabla u) + \lambda \partial_1 u\right] \cdot \mathcal{P}_\mathrm{H}\Delta u^* \, dx.$$

Taking the real parts of both sides, observing that

$$\operatorname{Re} \int_{\Omega_R} iku \cdot \mathcal{P}_\mathrm{H}\Delta u^* \, dx = -\operatorname{Re}\left(ik\|\nabla u\|_2^2\right) = 0,$$

and using Hölder's inequality, we conclude the estimate

$$\|\mathcal{P}_\mathrm{H}\Delta u\|_2^2 \leq \left(\|F\|_2 + \lambda\|\partial_1 u\|_2\right)\|\mathcal{P}_\mathrm{H}\Delta u\|_2$$
$$+ \operatorname{Re} \int_{\Omega_R} \omega(e_1 \wedge u - e_1 \wedge x \cdot \nabla u) \cdot \mathcal{P}_\mathrm{H}\Delta u^* \, dx. \tag{4.38}$$

In view of the identity from Lemma 4.1.9, we can estimate the remaining integral on the right-hand side to obtain

$$\text{Re} \int_{\Omega_R} \omega(e_1 \wedge u - e_1 \wedge x \cdot \nabla u) \cdot \mathcal{P}_H \Delta u^* \, dx \le c_2 \big(\|\nabla u\|_{2;\partial\Omega}^2 + \|\nabla u\|_{2;\Omega_R}^2 \big)$$

with c_2 independent of R. Employing the trace inequality from Theorem 2.3.2 on the domain Ω_{R_0}, we further estimate

$$\text{Re} \int_{\Omega_R} \omega(e_1 \wedge u - e_1 \wedge x \cdot \nabla u) \cdot \mathcal{P}_H \Delta u^* \, dx \le c_3 \|\nabla u\|_{2;\Omega_R}^2 + \varepsilon \|\nabla^2 u\|_{2;\Omega_R}^2$$

for c_3 dependent on $\varepsilon > 0$ but independent of R. We estimate the second term on the right-hand side with the help of Lemma 2.4.7 and deduce

$$\text{Re} \int_{\Omega_R} \omega(e_1 \wedge u - e_1 \wedge x \cdot \nabla u) \cdot \mathcal{P}_H \Delta u^* \, dx$$

$$\le (c_3 + c_4) \|\nabla u\|_{2;\Omega_R}^2 + \varepsilon c_4 \|\mathcal{P}_H \Delta u\|_{2;\Omega_R}^2$$

with a constant $c_4 > 0$ independent of R and ε. Combining this estimate with (4.38) and choosing ε sufficiently small, we obtain

$$\|\mathcal{P}_H \Delta u\|_2^2 \le c_5 \big(\|F\|_2 + \|\nabla u\|_2 \big) \|\mathcal{P}_H \Delta u\|_2 + c_6 \|\nabla u\|_2^2.$$

Employing Young's inequality and estimate (4.37), we arrive at

$$\|\mathcal{P}_H \Delta u\|_{2;\Omega_R} \le c_7 \big(\|F\|_2 + \|F\|_{6/5} \big).$$

Using Lemma 2.4.7 and estimate (4.37) once again and restoring the original notation, we end up with

$$\|\nabla^2 u_n\|_{2;\Omega_R} \le c_8 \big(\|F\|_2 + \|F\|_{6/5} \big) \tag{4.39}$$

with c_8 independent of R.

In particular, we see from (4.37), (4.39) and Poincaré's inequality that (u_n) is uniformly bounded in $W^{2,2}(\Omega_R)$ and thus contains a subsequence that converges weakly to some function $v \in L_\sigma^2(\Omega_R) \cap W_0^{1,2}(\Omega_R) \cap W^{2,2}(\Omega_R)$, which obeys the estimate

$$\|v\|_{6;\Omega_R} + \|\nabla v\|_{1,2;\Omega_R} \le c_9 \big(\|F\|_{6/5} + \|F\|_2 \big) \tag{4.40}$$

with c_9 independent of R. Moreover, since (ψ_j) is a basis of $L_\sigma^2(\Omega_R)$ and v satisfies (4.35) for all $j \in \mathbb{N}$, we obtain the identity

$$\mathcal{P}_H \big(\omega(ikv + e_1 \wedge v - e_1 \wedge x \cdot \nabla v) - \Delta v - \lambda \partial_1 v \big) = \mathcal{P}_H F.$$

By the Helmholtz–Weyl decomposition (Theorem 2.4.1), there exists $p \in W^{1,2}(\Omega_R)$ such that

$$\omega(ikv + e_1 \wedge v - e_1 \wedge x \cdot \nabla v) - \Delta v - \lambda \partial_1 v + \nabla p = F \qquad \text{in } \Omega_R.$$

In particular, since $v \in L^2_\sigma(\Omega_R) \cap W^{1,2}_0(\Omega_R)$, the functions v and p satisfy (4.33). Since $e_1 \wedge v - e_1 \wedge x \cdot \nabla v \in L^2_\sigma(\Omega_R)$ by Lemma 4.1.9, from this equation and (4.40) we further deduce the estimate

$$\omega \| ikv + e_1 \wedge v - e_1 \wedge x \cdot \nabla v \|_2 = \omega \big\| \mathcal{P}_H (ikv + e_1 \wedge v - e_1 \wedge x \cdot \nabla v) \big\|_2$$
$$\leq \| \mathcal{P}_H F \|_2 + \| \mathcal{P}_H \Delta v \|_2 + \lambda \| \mathcal{P}_H \partial_1 v \|_2 \leq c_{10} \big(\| F \|_{6/5} + \| F \|_2 \big)$$

by continuity of the Helmholtz projector. Combining this estimate with (4.40), we conclude (4.34). $\qquad\square$

In order to obtain a solution to the resolvent problem (4.4) in an exterior domain, the idea is to pass to the limit $R \to \infty$ in Lemma 4.1.12. To make this approach rigorous, we first have to employ a suitable cut-off argument.

Lemma 4.1.13. *Let Ω, λ, ω, k, F be as in Lemma 4.1.12. Then there exists a solution (v,p) to (4.4) with*

$$ikv + e_1 \wedge v - e_1 \wedge x \cdot \nabla v, \ \nabla v, \ \nabla^2 v, \ \nabla p \in L^2(\Omega), \qquad v \in L^6(\Omega).$$

Proof. Consider a standard cut-off function $\chi \in C_0^\infty(\mathbb{R};[0,1])$ with $\chi(s) = 1$ for $|s| \leq 1/2$ and $\chi(s) = 0$ for $|s| \geq 3/4$. For $m \in \mathbb{N}$ with $2m > \delta(\Omega^c)$ define $\chi_m \in C_0^\infty(\mathbb{R}^n)$ by $\chi_m(x) := \chi(|x|/m)$. Then we have

$$\chi_m(x) = 1 \text{ for } |x| \leq \frac{m}{2}, \quad \chi_m(x) = 0 \text{ for } |x| \geq \frac{3m}{4},$$
$$|\nabla \chi_m| \leq \frac{c_0}{m}, \qquad\qquad |\nabla^2 \chi_m| \leq \frac{c_1}{m^2},$$

where the last estimate follows from $\operatorname{supp} \nabla \chi_m \subset A_m := B_m \setminus B_{m/2}$. Define $w^m := \chi_m v^m$, where $v^m = v^R$ is the velocity field from Lemma 4.1.12 with $R = m$. Then w^m is an element of $W^{2,2}(\Omega)$, and $\|w^m\|_6 \leq \|v^m\|_6$. Hölder's inequality further yields

$$\|\nabla w^m\|_2 \leq c_2 \big(\|\nabla v^m\|_2 \|\chi\|_\infty + \|v^m\|_6 \|\nabla \chi_m\|_{3;A_m} \big)$$
$$\leq c_3 \big(\|\nabla v^m\|_2 + \|v^m\|_6 \big),$$
$$\|\nabla^2 w^m\|_2 \leq c_4 \big(\|\nabla^2 v^m\|_2 \|\chi_m\|_\infty + \|\nabla v^m\|_2 \|\nabla \chi_m\|_\infty + \|v^m\|_6 \|\nabla^2 \chi_m\|_{3;A_m} \big)$$
$$\leq c_5 \big(\|\nabla^2 v^m\|_2 + \|\nabla v^m\|_2 + \|v^m\|_6 \big).$$

Moreover, by $\nabla \chi_m(x) = \chi'(|x|/m)\frac{x}{m|x|}$, we have

$$e_1 \wedge x \cdot \nabla w^m = e_1 \wedge x \cdot \nabla v^m \chi_m + e_1 \wedge x \cdot \nabla \chi_m v^m = e_1 \wedge x \cdot \nabla v^m \chi_m,$$

from which we conclude

$$\|ikw^m + e_1 \wedge w^m - e_1 \wedge x \cdot \nabla w^m\|_2 \le \|ikv^m + e_1 \wedge v^m - e_1 \wedge x \cdot \nabla v^m\|_2.$$

Combining the above estimates with (4.34), we deduce

$$\|w^m\|_6 + \|\nabla w^m\|_2 + \|\nabla^2 w^m\|_2 + \omega\|ikw^m + e_1 \wedge w^m - e_1 \wedge x \cdot \nabla w^m\|_2$$
$$\le c_6\big(\|F\|_{6/5} + \|F\|_2\big)$$

with c_6 independent of m. This implies the existence of a subsequence of (w^m), still denoted by (w^m), that converges in the sense of distributions to some function $v \in W^{2,2}_{\mathrm{loc}}(\Omega)$ that satisfies

$$\|v\|_6 + \|\nabla v\|_2 + \|\nabla^2 v\|_2 + \omega\|ikv + e_1 \wedge v - e_1 \wedge x \cdot \nabla v\|_2 \le C_{39}\big(\|F\|_{6/5} + \|F\|_2\big).$$

Moreover, $v|_{\partial\Omega} = 0$. Let $\varphi \in C^\infty_0(\Omega)$ and choose $m_0 \in \mathbb{N}$ such that $\mathrm{supp}\,\varphi$ is contained in $\Omega_{m_0/2}$. For $m \ge m_0$ we have $w^m = v^m$ on $\Omega_{m_0/2}$ and thus

$$\int_\Omega w^m \cdot \nabla \varphi \, dx = \int_\Omega v^m \cdot \nabla \varphi \, dx = 0$$

by $(4.33)_2$. Passing to the limit $m \to \infty$, we conclude $\mathrm{div}\, v = 0$. Now let $\psi \in C^\infty_{0,\sigma}(\Omega)$ and choose m_0 such that $\mathrm{supp}\,\psi \subset \Omega_{m_0/2}$. With the same argument as above, for $m \ge m_0$ we obtain from $(4.33)_1$ that

$$\int_\Omega \big(\omega(ikw^m + e_1 \wedge w^m - e_1 \wedge x \cdot \nabla w^m) - \Delta w^m - \lambda\partial_1 w^m - F\big) \cdot \psi \, dx$$
$$= \int_\Omega \big(\omega(ikv^m + e_1 \wedge v^m - e_1 \wedge x \cdot \nabla v^m) - \Delta v^m - \lambda\partial_1 v^m - F\big) \cdot \psi \, dx = 0.$$

Therefore, by passing to the limit $m \to \infty$, we see

$$\int_\Omega \big(\omega(ikv + e_1 \wedge v - e_1 \wedge x \cdot \nabla v) - \Delta v - \lambda\partial_1 v - F\big) \cdot \psi \, dx = 0$$

for all $\psi \in C^\infty_{0,\sigma}(\Omega)$. Consequently, by the Helmholtz–Weyl decomposition in $L^2(\Omega)$, there exists a function $p \in D^{1,2}(\Omega)$ such that (v, p) is a solution to (4.4). This completes the proof. $\qquad\square$

4.1.5 Well-Posedness

Combining Lemma 4.1.6, Lemma 4.1.8 and Lemma 4.1.13, we conclude the following well-posedness result for the resolvent problem (4.4). For the proof we first consider smooth data F, for which we obtain a solution in an L^2 framework by Lemma 4.1.13. Subsequently, we show by another cut-off argument that this solution belongs to the function space defined by the left-hand side of the resolvent estimate (4.25). Finally, a classical density argument yields a solution to (4.4) for general data $F \in L^q(\Omega)$. In total, we obtain the following result.

Theorem 4.1.14. *Let $\Omega \subset \mathbb{R}^3$ be an exterior domain of class C^3. Let $q \in (1,2)$, $k \in \mathbb{Z}$ and $\lambda, \omega, \theta, B > 0$ with $\lambda^2 \leq \theta\omega \leq B$. For every $F \in L^q(\Omega)^3$ there exists a solution $(v,p) \in W^{2,q}_{loc}(\overline{\Omega})^3 \times W^{1,q}_{loc}(\overline{\Omega})$ to the resolvent problem (4.4) subject to the estimate*

$$\omega\|ikv + e_1 \wedge v - e_1 \wedge x \cdot \nabla v\|_q + \|\nabla^2 v\|_q + \lambda\|\partial_1 v\|_q$$
$$+ \lambda^{1/2}\|v\|_{s_1} + \lambda^{1/4}\|\nabla v\|_{s_2} + \|\nabla p\|_q \leq C_{40}\|F\|_q \tag{4.41}$$

for a constant $C_{40} = C_{40}(\Omega, q, \lambda, \omega) > 0$ and $s_1 = 2q/(2-q)$, $s_2 = 4q/(4-q)$.

Additionally, if $(w, \mathfrak{q})$ is another solution to (4.4) in the function class defined by the norms on the left-hand side of (4.41), then $v = w$, and $p - \mathfrak{q}$ is a constant.

Moreover, if $q \in (1, \frac{3}{2})$, then the constant C_{40} can be chosen independently of λ and ω such that $C_{40} = C_{40}(\Omega, q, \theta, B)$.

Proof. The uniqueness statement is an immediate consequence of Lemma 4.1.6, and estimate (4.41) has been proved in Lemma 4.1.8. It thus remains to show existence of a solution for $F \in L^q(\Omega)$.

At first, let $F \in C_0^\infty(\Omega)$, and let (v,p) be the corresponding solution to (4.4) that exists by Lemma 4.1.13. We first show that then (v,p) belongs to the correct function class. By Hölder's inequality, we directly find that

$$v \in W^{2,q}(\Omega_\rho), \qquad p \in W^{1,q}(\Omega_\rho) \tag{4.42}$$

for any $\rho > R$ and all $q \in [1,2]$. Repeating the cut-off argument from (4.15), we obtain a solution $(u, \mathfrak{p})$ to (4.16) for some function $f \in L^2(\mathbb{R}^3)$ with compact support. In particular, this implies $f \in L^q(\mathbb{R}^3)$ for all $q \in (1,2)$. Theorem 4.1.5 yields existence of a solution to (4.16) satisfying (4.12). Since $u \in L^6(\mathbb{R}^3)$, Theorem 4.1.5 further ensures that $(u, \mathfrak{p})$ coincides with this solution up to an additive constant for $\mathfrak{p}$. We thus have

$$iku + e_1 \wedge u - e_1 \wedge x \cdot \nabla u, \ \nabla^2 u, \ \partial_1 u, \ \nabla\mathfrak{p} \in L^q(\mathbb{R}^3),$$
$$u \in L^{2q/(2-q)}(\mathbb{R}^3), \qquad \nabla u \in L^{4q/(4-q)}(\mathbb{R}^3).$$

Since $v = u$ and $p = \mathfrak{p}$ on B^{2R}, the integrability properties above in combination with (4.42) show that v and p belong to the correct function spaces.

Now consider general $F \in \mathrm{L}^q(\Omega)$ and a sequence $(F_j) \subset \mathrm{C}_0^\infty(\Omega)$ that converges to F in $\mathrm{L}^q(\Omega)$. As seen above, for each $j \in \mathbb{N}$ there exists a solution $(v, p) = (v_j, p_j)$ to (4.4) with $F = F_j$, which obeys estimate (4.41). Additionally, this implies that $(v_j, \nabla p_j)$ is a Cauchy sequence in the function space defined by the norm on the left-hand side of (4.41), and thus possesses a limit $(v, \nabla p)$, which satisfies the resolvent problem (4.4) and the resolvent estimate (4.41). $\qquad\qquad\square$

Remark 4.1.15. Note that for $k = 0$ we recover the well-known L^q theory for the corresponding stationary problem; see [42, Theorem VIII.8.1] for example.

Remark 4.1.16. In the classical study of resolvent problems, one considers an operator A in a Banach space X and investigates solvability of the equation $su - Au = f$ for $s \in \mathbb{C}$ together with a corresponding resolvent estimate of the form

$$|s| \, \|u\|_X + \|Au\|_X \leq C_{41} \|f\|_X.$$

In particular, if $s \neq 0$, the solution u belongs to the same space X as the data f. In contrast, the solution theory established in Theorem 4.1.14 does not fit into this setting, and we do not obtain a classical resolvent estimate.

4.2 The Time-Periodic Linear Problem

After having established well-posedness of the resolvent problem in Theorem 4.1.14, we now turn to the time-periodic problem

$$\begin{cases} \omega(\partial_t u + e_1 \wedge u - e_1 \wedge x \cdot \nabla u) - \Delta u - \lambda \partial_1 u + \nabla \mathfrak{p} = f & \text{in } \mathbb{T} \times \Omega, \\ \operatorname{div} u = 0 & \text{in } \mathbb{T} \times \Omega, \\ u = 0 & \text{on } \mathbb{T} \times \partial\Omega, \end{cases} \quad (4.43)$$

where $\mathbb{T} = \mathbb{R}/2\pi\mathbb{Z}$. In order to transfer the properties of the resolvent problem to the time-periodic case, we work in a framework of absolutely convergent Fourier series. Besides uniqueness and existence of a solution, this allows to obtain a corresponding *a priori* estimate with the same constant as in the resolvent estimate (4.41). In conclusion, we obtain conditions when the constant is independent of λ and ω, which is crucial for the fixed-point argument employed to solve the nonlinear problem (4.1).

We first introduce the spaces of absolutely convergent Fourier series and provide some inequalities necessary for the treatment of the nonlinear problem (4.1). Afterwards, we establish well-posedness of the time-periodic problem (4.43) in these function spaces.

4.2.1 The Functional Framework

Here we introduce the functional framework where we carry out the analysis of the time-periodic problem (4.1). For simplicity, we restrict to the case of 2π-periodic functions, that is, to the torus group $\mathbb{T} = \mathbb{R}/2\pi\mathbb{Z}$.

Let X be a semi-normed vector space with semi-norm $\|\cdot\|_X$. To any sequence $(f_k)_{k\in\mathbb{Z}} \subset X$ we can formally associate a Fourier series

$$f(t) := \sum_{k\in\mathbb{Z}} f_k\, e^{ikt}. \tag{4.44}$$

This series converges pointwise if the sequence of norms $(\|f_k\|_X)_{k\in\mathbb{Z}}$ is an element of $\ell^1(\mathbb{Z};\mathbb{R})$, which means $(f_k)_{k\in\mathbb{Z}} \in \ell^1(\mathbb{Z};X)$. Recalling the inverse Fourier transformation $\mathscr{F}_{\mathbb{T}}^{-1}$ on $\mathbb{T}$, the equation (4.44) can thus be expressed as $f = \mathscr{F}_{\mathbb{T}}^{-1}[(f_k)_{k\in\mathbb{Z}}]$, so that f is an absolutely convergent Fourier series. Clearly, these objects constitute a function space defined by

$$A(\mathbb{T};X) := \left\{ f:\mathbb{T} \to X \;\middle|\; f(t) = \sum_{k\in\mathbb{Z}} f_k\, e^{ikt},\; (f_k)_{k\in\mathbb{Z}} \subset X,\; \sum_{k\in\mathbb{Z}} \|f_k\|_X < \infty \right\},$$

the space of functions on $\mathbb{T}$ with absolutely convergent Fourier series, which we equip with the semi-norm

$$\|f\|_{A(\mathbb{T};X)} := \sum_{k\in\mathbb{Z}} \|f_k\|_X.$$

Then $A(\mathbb{T};X)$ coincides with the space $\mathscr{F}_{\mathbb{T}}^{-1}[\ell^1(\mathbb{Z};X)]$, which embeds into the space $C(\mathbb{T};X)$ of X-valued continuous functions on $\mathbb{T}$. This property also justifies the pointwise definition in (4.44).

Observe that if X is a normed space, then $A(\mathbb{T};X)$ is also a normed space, and if X is a Banach space, then $A(\mathbb{T};X)$ is also a Banach space.

It is well known that the scalar-valued space $A(\mathbb{T};\mathbb{R})$ is a Banach algebra with respect to pointwise multiplication, the so-called Wiener algebra.

One can generalize this property and derive estimates in the X-valued case. For example, one readily shows the following correspondency of Hölder's inequality.

Proposition 4.2.1. *Let $D \subset \mathbb{R}^n$, $n \in \mathbb{N}$, be an open set and $p, q, r \in [1, \infty]$ such that $1/p + 1/q = 1/r$, whereby $1/\infty := 0$ as customary. Moreover, let $f \in A(\mathbb{T}; L^p(D))$ and $g \in A(\mathbb{T}; L^q(D))$. Then their product fg satisfies $fg \in A(\mathbb{T}; L^r(D))$ and*

$$\|fg\|_{A(\mathbb{T};L^r(D))} \leq \|f\|_{A(\mathbb{T};L^p(D))} \|g\|_{A(\mathbb{T};L^q(D))}. \tag{4.45}$$

Proof. By assumption we have $f = \mathscr{F}_{\mathbb{T}}^{-1}[(f_k)]$ and $g = \mathscr{F}_{\mathbb{T}}^{-1}[(g_k)]$ for elements $(f_k) \in \ell^1(\mathbb{Z}; L^p(D))$ and $(g_k) \in \ell^1(\mathbb{Z}; L^q(D))$. Therefore, we have $fg = \mathscr{F}_{\mathbb{T}}^{-1}[(f_k) *_{\mathbb{Z}} (g_k)]$. Then the classical Hölder inequality implies

$$\|fg\|_{A(\mathbb{T};L^r(D))} = \sum_{k \in \mathbb{Z}} \left\| \sum_{\ell \in \mathbb{Z}} f_\ell g_{k-\ell} \right\|_{L^r(D)} \leq \sum_{k \in \mathbb{Z}} \sum_{\ell \in \mathbb{Z}} \|f_\ell g_{k-\ell}\|_{L^r(D)}$$

$$\leq \sum_{k \in \mathbb{Z}} \sum_{\ell \in \mathbb{Z}} \|f_\ell\|_{L^p(D)} \|g_{k-\ell}\|_{L^q(D)} = \|f\|_{A(\mathbb{T};L^p(D))} \|g\|_{A(\mathbb{T};L^q(D))}.$$

This completes the proof. □

Remark 4.2.2. Here we single out the special case when one of the functions involved in Proposition 4.2.1 only depends on time. Let $f \in A(\mathbb{T}; \mathbb{R})$ and $g \in A(\mathbb{T}; L^q(\Omega))$. Then, the function f can also be regarded as an element of $A(\mathbb{T}; L^\infty(D))$. Consequently, (4.45) implies $fg \in A(\mathbb{T}; L^q(D))$ and

$$\|fg\|_{A(\mathbb{T};L^q(D))} \leq \|f\|_{A(\mathbb{T};\mathbb{R})} \|g\|_{A(\mathbb{T};L^q(D))}.$$

Similarly to Proposition 4.2.1, we transfer the classical interpolation inequality for Lebesgue spaces to the framework of the spaces $A(\mathbb{T}; X)$.

Proposition 4.2.3. *Let $D \subset \mathbb{R}^n$, $n \in \mathbb{N}$, be an open set and $p, q, r \in [1, \infty]$ such that*

$$\frac{1-\theta}{p} + \frac{\theta}{q} = \frac{1}{r}$$

for some $\theta \in [0, 1]$, whereby $1/\infty := 0$ as customary. Then every function $f \in A(\mathbb{T}; L^p(D)) \cap A(\mathbb{T}; L^q(D))$ satisfies $f \in A(\mathbb{T}; L^r(D))$ and

$$\|f\|_{A(\mathbb{T};L^r(D))} \leq \|f\|_{A(\mathbb{T};L^p(D))}^{1-\theta} \|f\|_{A(\mathbb{T};L^q(D))}^{\theta}. \tag{4.46}$$

Proof. By assumption, we have the identity $f = \mathscr{F}_{\mathbb{T}}^{-1}[(f_k)]$ for an element $(f_k) \in \ell^1(\mathbb{Z}; L^p(D) \cap L^q(D))$. The classical interpolation inequality for Lebesgue spaces together with an application of Hölder's inequality on $\mathbb{Z}$ yields

$$\|f\|_{A(\mathbb{T};L^r(D))} = \sum_{k\in\mathbb{Z}} \|f_k\|_{L^r(D)} \leq \sum_{k\in\mathbb{Z}} \|f_k\|_{L^p(D)}^{1-\theta} \|f_k\|_{L^q(D)}^{\theta}$$

$$\leq \|f\|_{A(\mathbb{T};L^p(D))}^{1-\theta} \|f\|_{A(\mathbb{T};L^q(D))}^{\theta}.$$

This completes the proof. $\qquad\square$

Observe that one can directly generalize Proposition 4.2.1 and Proposition 4.2.3 to general measure spaces. For the sake of simplicity, we have restricted ourselves to the Lebesgue measure here.

Remark 4.2.4. Note that we can easily transfer embedding properties to the framework of these spaces. Let X and Y be two semi-normed spaces, equipped with the semi-norms $\|\cdot\|_X$ and $\|\cdot\|_Y$, such that $X \hookrightarrow Y$, that is, X is continuously embedded in Y. This directly implies $A(\mathbb{T}; X) \hookrightarrow A(\mathbb{T}; Y)$ since for $f \in A(\mathbb{T}; X)$ with Fourier coefficients $(f_k)_{k\in\mathbb{Z}} \in \ell^1(\mathbb{Z}; X)$ one has

$$\sum_{k\in\mathbb{Z}} \|f_k\|_Y \leq C_{42} \sum_{k\in\mathbb{Z}} \|f_k\|_X = C_{42} \|f\|_{A(\mathbb{T};X)},$$

where C_{42} is the embedding constant of the embedding $X \hookrightarrow Y$.

The main advantage of the space $A(\mathbb{T}; X)$ is that we can directly transfer *a priori* estimates for a resolvent problem to *a priori* estimates for the corresponding time-periodic problem. Recall that in general this is not possible in a classical Lebesgue framework of spaces of the form $L^q(\mathbb{T}; X)$ as we pointed out before.

4.2.2 Well-Posedness

Now we show the existence of a solution to the time-periodic problem (4.43) by reducing it to the resolvent problem (4.4) treated in Theorem 4.1.14.

Theorem 4.2.5. *Let $\Omega \subset \mathbb{R}^3$ be an exterior domain of class C^3. Let $q \in (1,2)$ and $\lambda, \omega, \theta, B > 0$ with $\lambda^2 \leq \theta\omega \leq B$. For every $f \in A(\mathbb{T}; L^q(\Omega))^3$ there exists a solution $(u, \mathfrak{p})$ to (4.43) subject to the estimate*

$$\omega\|\partial_t u + e_1 \wedge u - e_1 \wedge x \cdot \nabla u\|_{A(\mathbb{T};L^q(\Omega))} + \|\nabla^2 u\|_{A(\mathbb{T};L^q(\Omega))}$$

$$+ \lambda\|\partial_1 u\|_{A(\mathbb{T};L^q(\Omega))} + \lambda^{1/2}\|u\|_{A(\mathbb{T};L^{s_1}(\Omega))} + \lambda^{1/4}\|\nabla u\|_{A(\mathbb{T};L^{s_2}(\Omega))} \tag{4.47}$$

$$+ \|\nabla\mathfrak{p}\|_{A(\mathbb{T};L^q(\Omega))} \leq C_{43}\|f\|_{A(\mathbb{T};L^q(\Omega))}$$

for a constant $C_{43} = C_{43}(\Omega, q, \lambda, \omega) > 0$ and $s_1 = 2q/(2-q)$, $s_2 = 4q/(4-q)$.

Additionally, if $(w, \mathfrak{q})$ is another solution to (4.43) in the function class defined by the norms on the left-hand side of (4.47), then $u = w$ and $\mathfrak{p} = \mathfrak{q} + \mathfrak{q}_0$ for some (spatially constant) function $\mathfrak{q}_0 \colon \mathbb{T} \to \mathbb{R}$.

Moreover, if $q \in (1, \frac{3}{2})$, then the constant C_{43} can be chosen independently of λ and ω such that $C_{43} = C_{43}(\Omega, q, \theta, B)$.

Proof. If $(u, \mathfrak{p})$ is a solution to (4.43), in the above function class, then the k-th Fourier coefficients $v := \mathscr{F}_{\mathbb{T}}^{-1}[u](k)$ and $p := \mathscr{F}_{\mathbb{T}}^{-1}[\mathfrak{p}](k)$ satisfy the resolvent problem (4.4) with right-hand side $F := \mathscr{F}_{\mathbb{T}}^{-1}[f](k)$. The uniqueness statement of Theorem 4.2.5 is thus a direct consequence of the uniqueness statement of Theorem 4.1.14.

To show existence of a solution satisfying (4.47), consider a function $f \in A(\mathbb{T}; L^q(\Omega))$. Then

$$f(t, x) = \sum_{k \in \mathbb{Z}} f_k(x) e^{ikt}$$

with $f_k \in L^q(\Omega)$ for $k \in \mathbb{Z}$. Let $(u_k, \mathfrak{p}_k) = (v, p)$ be a solution to the resolvent problem (4.4) with right-hand side $F = f_k$ that exists due to Theorem 4.1.14. We define

$$u(t, x) := \sum_{k \in \mathbb{Z}} u_k(x) e^{ikt}, \qquad \mathfrak{p}(t, x) := \sum_{k \in \mathbb{Z}} \mathfrak{p}_k(x) e^{ikt}.$$

By (4.41), u and $\mathfrak{p}$ are well defined and satisfy problem (4.43). We directly conclude estimate (4.47) from estimate (4.41) with coincident constants $C_{43} = C_{40}$. $\qquad \square$

Remark 4.2.6. Recall that the constant C_{40} in the resolvent estimate (4.41) was independent of the actual resolvent parameter $k \in \mathbb{Z}$. Therefore, as shown in the previous proof, it coincides with the constant C_{43} in the *a priori* estimate (4.47) for the time-periodic problem. In particular, we can thus choose the constant C_{43} independently of λ and ω if $1 < q < 3/2$, which will be crucial to show existence of a solution to (4.1) in the next section.

4.3 The Nonlinear Problem

Now we begin with the study of the nonlinear problem (4.1). First of all, we formulate it as a problem with homogeneous boundary conditions and derive suitable nonlinear estimates. We then apply the linear theory established in Theorem 4.2.5 to show existence of a solution to the nonlinear system by the contraction mapping principle.

4.3.1 Reformulation of the Problem

In order to show existence of a solution to (4.1), we transform it to a problem with homogeneous boundary conditions. To this end, we use the following simple lemma that is an analogue of Lemma 3.3.1.

Lemma 4.3.1. *Let $\Omega \subset \mathbb{R}^n$ be an exterior domain and $R > \delta(\Omega^c)$. Then there exists a function $W \in C_0^\infty(\mathbb{R}^n)^n$ with $\operatorname{supp} W \subset B_R$, $\operatorname{div} W \equiv 0$ and $W|_{\partial\Omega} = e_1 \wedge x$.*

Proof. The proof is analogous to that of Lemma 3.3.1. Consider a second radius $R_0 > 0$ with $\delta(\Omega^c) < R_0 < R$, and let $\varphi \in C_0^\infty(\mathbb{R}^n)$ with $\varphi \equiv 1$ on B_{R_0} and $\varphi \equiv 0$ on B^R. We define the function $W : \mathbb{R}^n \to \mathbb{R}^n$ by

$$W(x) := -\frac{1}{2}\operatorname{curl}\left[e_1 |x|^2 \varphi(x)\right]$$

Then $W \in C_0^\infty(\mathbb{R}^n)^n$, $\operatorname{supp} W \subset B_R$ and

$$\operatorname{div} W(x) = -\frac{1}{2}\operatorname{div}\operatorname{curl}\left[e_1 |x|^2 \varphi(x)\right] = 0$$

Moreover, for $x \in \partial\Omega$ we have $|x| < R_0$, and therefore

$$W(x) = -\frac{1}{2}\operatorname{curl}\left[e_1 |x|^2\right] = e_1 \wedge x.$$

This completes the proof. □

For fixed $R > \delta(\Omega^c)$, consider the functions V and W from Lemma 3.3.1 and Lemma 4.3.1 and define

$$U : \mathbb{T} \times \mathbb{R}^3 \to \mathbb{R}^3, \qquad U(t,x) = \alpha(t)V + \omega W. \tag{4.48}$$

Then $U(t, \cdot) \in C_0^\infty(\mathbb{R}^3)$ for all $t \in \mathbb{T}$, and the time regularity of U coincides with that of α. Below we will assume $\alpha, \frac{d}{dt}\alpha \in A(\mathbb{T}; \mathbb{R})$, so that $U \in C^1(\mathbb{T} \times \mathbb{R}^3)$. Now define $v := u - U$ and $p := \mathfrak{p}$. Then $(u, \mathfrak{p})$ solves (4.1) if and only if (v, p) solves

$$\begin{cases} \omega(\partial_t v + e_1 \wedge v - e_1 \wedge x \cdot \nabla v) - \Delta v - \lambda\partial_1 v + \nabla p = f + \mathcal{N}(v) & \text{in } \mathbb{T} \times \Omega, \\ \operatorname{div} v = 0 & \text{in } \mathbb{T} \times \Omega, \\ v = 0 & \text{on } \mathbb{T} \times \partial\Omega, \\ \lim_{|x| \to \infty} v(t,x) = 0 & \text{for } t \in \mathbb{T}, \end{cases}$$

$$\tag{4.49}$$

where

$$\mathcal{N}(v) \coloneqq (\mathcal{P}_\perp \alpha)\partial_1 v - \omega(\partial_t U + e_1 \wedge U - e_1 \wedge x \cdot \nabla U)$$
$$+ \Delta U + \alpha \partial_1 U - v \cdot \nabla v - U \cdot \nabla v - v \cdot \nabla U - U \cdot \nabla U. \tag{4.50}$$

Recall that $\mathcal{P}_\perp \alpha = \alpha - \lambda$. It thus remains to show existence of a solution to the nonlinear system (4.49).

4.3.2 Existence of a Solution

To show existence of a solution to (4.49), we first set up the functional framework. For the sake of clarity, we introduce the abbreviations

$$\|\alpha\|_{\mathrm{A}} \coloneqq \|\alpha\|_{\mathrm{A}(\mathbb{T};\mathbb{R})}, \qquad \|f\|_{\mathrm{A}^s} \coloneqq \|f\|_{\mathrm{A}(\mathbb{T};\mathrm{L}^s(\Omega))}$$

for $s \in (1, \infty)$. For $\lambda, \omega > 0$ and $q \in (1, 2)$ we define the function space

$$\mathcal{X}^q_{\lambda,\omega}(\mathbb{T} \times \Omega) \coloneqq \{v \in \mathrm{L}^1_{\mathrm{loc}}(\mathbb{T} \times \Omega) \mid \|v\|_{\mathcal{X}^q_{\lambda,\omega}} < \infty\},$$

where

$$\|v\|_{\mathcal{X}^q_{\lambda,\omega}} \coloneqq \omega\|\partial_t v + e_1 \wedge v - e_1 \wedge x \cdot \nabla v\|_{\mathrm{A}^q}$$
$$+ \|\nabla^2 v\|_{\mathrm{A}^q} + \lambda\|\partial_1 v\|_{\mathrm{A}^q} + \lambda^{1/2}\|v\|_{\mathrm{A}^{s_1}} + \lambda^{1/4}\|\nabla v\|_{\mathrm{A}^{s_2}}$$

with $s_1 \coloneqq 2q/(2-q)$ and $s_2 \coloneqq 4q/(4-q)$. We establish the following estimates of the nonlinear term $\mathcal{N}(v)$ from (4.50) under the assumption that the (time-dependent) translation velocity α satisfies $\alpha, \frac{\mathrm{d}}{\mathrm{d}t}\alpha \in \mathrm{A}(\mathbb{T};\mathbb{R})$.

Lemma 4.3.2. *Let* $q \in [\frac{6}{5}, \frac{4}{3}]$, $0 < \lambda \le \lambda_0$ *and* $0 < \omega \le \omega_0$. *Let* $v_1, v_2 \in \mathcal{X}^q_{\lambda,\omega}(\mathbb{T} \times \Omega)$. *There exists a constant* $C_{44} = C_{44}(\Omega, q, \lambda_0, \omega_0) > 0$ *such that*

$$\|\mathcal{N}(v_1)\|_{\mathrm{A}^q}$$
$$\le C_{44}\Big(\lambda^{-1}\|\mathcal{P}_\perp \alpha\|_{\mathrm{A}}\|v_1\|_{\mathcal{X}^q_{\lambda,\omega}} + \lambda^{-(3q-3)/q}\|v_1\|^2_{\mathcal{X}^q_{\lambda,\omega}} \tag{4.51}$$
$$+ (\lambda + \omega + \|\mathcal{P}_\perp \alpha\|_{\mathrm{A}})(1 + \|\mathcal{P}_\perp \alpha\|_{\mathrm{A}} + \|\tfrac{\mathrm{d}}{\mathrm{d}t}\alpha\|_{\mathrm{A}} + \|v_1\|_{\mathcal{X}^q_{\lambda,\omega}})\Big),$$
$$\|\mathcal{N}(v_1) - \mathcal{N}(v_2)\|_{\mathrm{A}^q}$$
$$\le C_{44}\Big(\lambda^{-1}\|\mathcal{P}_\perp \alpha\|_{\mathrm{A}} + \lambda + \omega + \|\mathcal{P}_\perp \alpha\|_{\mathrm{A}} \tag{4.52}$$
$$+ \lambda^{-(3q-3)/q}(\|v_1\|_{\mathcal{X}^q_{\lambda,\omega}} + \|v_2\|_{\mathcal{X}^q_{\lambda,\omega}})\Big)\|v_1 - v_2\|_{\mathcal{X}^q_{\lambda,\omega}}.$$

Proof. Let $v \in \mathcal{X}^q_{\lambda,\omega}(\mathbb{T} \times \Omega)$. In view of Remark 4.2.2, we have

$$\|(\mathcal{P}_\perp \alpha)\partial_1 v\|_{\mathrm{A}^q} \le \|\mathcal{P}_\perp \alpha\|_{\mathrm{A}}\|\partial_1 v\|_{\mathrm{A}^q} \le \|\mathcal{P}_\perp \alpha\|_{\mathrm{A}}\lambda^{-1}\|v\|_{\mathcal{X}^q_{\lambda,\omega}}.$$

By the definition of U in (4.48) and the decomposition $\alpha = \lambda + \mathcal{P}_\perp \alpha$, we further have

$$\omega \|\partial_t U + e_1 \wedge U - e_1 \wedge x \cdot \nabla U\|_{A^q} \le c_0 \omega \left(\|\tfrac{d}{dt}\alpha\|_A \|V\|_{A^q} + \|U\|_{A^q} + \|\nabla U\|_{A^q} \right)$$
$$\le c_1 \omega \left(\|\tfrac{d}{dt}\alpha\|_A + \lambda + \omega + \|\mathcal{P}_\perp \alpha\|_A \right).$$

Similarly, we obtain

$$\|\Delta U\|_{A^q} \le c_2 \left(\lambda + \omega + \|\mathcal{P}_\perp \alpha\|_A \right),$$
$$\|\alpha \partial_1 U\|_{A^q} \le c_3 \left(\lambda + \|\mathcal{P}_\perp \alpha\|_A \right) \left(\lambda + \omega + \|\mathcal{P}_\perp \alpha\|_A \right).$$

Additionally using Proposition 4.2.1, the Sobolev inequality and Remark 4.2.4, we further have

$$\|U \cdot \nabla v + v \cdot \nabla U + U \cdot \nabla U\|_{A^q}$$
$$\le c_4 \left(\|v\|_{A^{3q/(3-2q)}} \|\nabla U\|_{A^{3/2}} + \|U\|_{A^3} \|\nabla v\|_{A^{3q/(3-q)}} + \|U\|_{A^{2q}} \|\nabla U\|_{A^{2q}} \right)$$
$$\le c_5 \left(\lambda + \omega + \|\mathcal{P}_\perp \alpha\|_A \right) \left(\|v\|_{\mathcal{X}^q_{\lambda,\omega}} + \lambda + \omega + \|\mathcal{P}_\perp \alpha\|_A \right).$$

Finally, let us treat the critical term $v \cdot \nabla v$. Since the assumption $q \in \left[\frac{6}{5}, \frac{4}{3}\right]$ implies $\frac{4q}{4-q} \le 2 \le \frac{3q}{3-q}$, we can employ estimates (4.45) and (4.46) to obtain

$$\|v \cdot \nabla v\|_{A^q} \le \|v\|_{A^{2q/(2-q)}} \|\nabla v\|_{A^2} \le c_6 \|v\|_{A^{2q/(2-q)}} \|\nabla v\|_{A^{4q/(4-q)}}^{1-\theta} \|\nabla v\|_{A^{3q/(3-q)}}^{\theta}$$

with $\theta = \frac{12-9q}{q}$. By the Sobolev inequality and Remark 4.2.4, we deduce

$$\|v \cdot \nabla v\|_{A^q} \le c_7 \lambda^{-1/2-(1-\theta)/4} \|v\|_{\mathcal{X}^q_{\lambda,\omega}}^{2-\theta} \|\nabla^2 v\|_{A^q}^{\theta} \le c_8 \lambda^{-(3q-3)/q} \|v\|_{\mathcal{X}^q_{\lambda,\omega}}^{2}.$$

Collecting the above estimates and using $\lambda \le \lambda_0$ and $\omega \le \omega_0$, we obtain (4.51). Estimate (4.52) follows in the same way. □

After these preparations, we can now show the existence of a solution to the nonlinear problem (4.1) by reformulating (4.49) as a fixed-point equation by means of the solution theory to the time-periodic problem established in Theorem 4.2.5. We then establish existence of a solution to the resulting fixed-point equation by the contraction mapping principle.

Theorem 4.3.3. *Let $\Omega \subset \mathbb{R}^3$ be an exterior domain of class C^3, and let $q \in \left[\frac{6}{5}, \frac{4}{3}\right]$. Let $f \in A(\mathbb{T}; L^q(\Omega))^3$ and $\alpha \in A(\mathbb{T}; \mathbb{R})$ such that $\frac{d}{dt}\alpha \in A(\mathbb{T}; \mathbb{R})$ and define*

$$\lambda := \int_{\mathbb{T}} \alpha(t)\, dt.$$

For all $\rho \in \left(\frac{3q-3}{q}, 1\right)$ and $\theta > 0$ there are constants $\kappa > 0$ and $\lambda_0 > 0$ such that for all

$$\lambda \in (0, \lambda_0], \qquad \omega \in \left[\frac{\lambda^2}{\theta}, \kappa\lambda^\rho\right] \tag{4.53}$$

there exists $\varepsilon > 0$ such that if

$$\|\alpha - \lambda\|_{A(\mathbb{T};\mathbb{R})} + \|f\|_{A(\mathbb{T};L^q(\Omega))} \le \varepsilon,$$

then there is a solution $(u, \mathfrak{p})$ to (4.1) with

$$u \in A(\mathbb{T}; L^{2q/(2-q)}(\Omega))^3,$$
$$\nabla u \in A(\mathbb{T}; L^{4q/(4-q)}(\Omega))^{3\times3},$$
$$\nabla^2 u \in A(\mathbb{T}; L^q(\Omega))^{3\times3\times3},$$
$$\partial_t u + e_1 \wedge u - e_1 \wedge x \cdot \nabla u, \ \partial_1 u, \ \nabla\mathfrak{p} \in A(\mathbb{T}; L^q(\Omega))^3.$$

Proof. In order to obtain a solution to (4.49) by a fixed-point argument, we consider the problem

$$\begin{cases} \omega(\partial_t w + e_1 \wedge w - e_1 \wedge x \cdot \nabla w) - \Delta w - \lambda\partial_1 w + \nabla\mathfrak{q} = f + \mathcal{N}(v) & \text{in } \mathbb{T}\times\Omega, \\ \operatorname{div} w = 0 & \text{in } \mathbb{T}\times\Omega, \\ w = 0 & \text{on } \mathbb{T}\times\partial\Omega, \end{cases} \tag{4.54}$$

for given $v \in \mathcal{X}^q_{\lambda,\omega}(\mathbb{T}\times\Omega)$. Since Lemma 4.3.2 ensures that $\mathcal{N}(v) \in A(\mathbb{T}; L^q(\Omega))$, Theorem 4.2.5 shows the existence of a unique velocity field $w \in \mathcal{X}^q_{\lambda,\omega}(\mathbb{T}\times\Omega)$ and a pressure field $\mathfrak{q}$ with $\nabla\mathfrak{q} \in A(\mathbb{T}; L^q(\Omega))^3$ that satisfy (4.54). We thereby obtain a solution map $\mathcal{S}_{\lambda,\omega}: \mathcal{X}^q_{\lambda,\omega} \to \mathcal{X}^q_{\lambda,\omega}$, $v \mapsto w$. Moreover, Theorem 4.2.5 (where we can set $B = 1$) and Lemma 4.3.2 yield the estimate

$$\begin{aligned} \|\mathcal{S}_{\lambda,\omega}(v)\|_{\mathcal{X}^q_{\lambda,\omega}} &\le C_{40}\big(\|f\|_{A^q} + \|\mathcal{N}(v)\|_{A^q}\big) \\ &\le c_0\big(\varepsilon + \varepsilon\lambda^{-1}\|v\|_{\mathcal{X}^q_{\lambda,\omega}} + \lambda^{-(3q-3)/q}\|v\|^2_{\mathcal{X}^q_{\lambda,\omega}} \\ &\quad + (\lambda + \omega + \varepsilon)\big(1 + \varepsilon + \|\tfrac{d}{dt}\alpha\|_{A(\mathbb{T};\mathbb{R})} + \|v\|_{\mathcal{X}^q_{\lambda,\omega}}\big)\big) \end{aligned}$$

with $c_0 = C_{40}(1 + C_{44})$ independent of λ and ω. Therefore, $\mathcal{S}_{\lambda,\omega}$ is a self-mapping on the closed subset

$$M_\delta := \big\{v \in \mathcal{X}^q_{\lambda,\omega} \mid \|v\|_{\mathcal{X}^q_{\lambda,\omega}} \le \delta\big\}$$

of $\mathcal{X}^q_{\lambda,\omega}(\mathbb{T} \times \Omega)$ provided

$$c_0\left(\varepsilon + \varepsilon\lambda^{-1}\delta + \lambda^{-(3q-3)/q}\delta^2 + (\lambda + \omega + \varepsilon)\left(1 + \varepsilon + \left\|\tfrac{\mathrm{d}}{\mathrm{d}t}\alpha\right\|_{A(\mathbb{T};\mathbb{R})} + \delta\right)\right) \leq \delta. \quad (4.55)$$

For $\rho \in \left(\tfrac{3q-3}{q}, 1\right)$ we choose $\delta := \lambda^\rho$. Moreover, let $\varepsilon = \lambda^2$ and $\omega \leq \kappa\delta$ for some $\kappa > 0$. Then (4.55) holds if

$$c_0\left(\lambda^{2-\rho} + \lambda + \lambda^{\rho-(3q-3)/q} + \left(\lambda^{1-\rho} + \kappa + \lambda^{2-\rho}\right)\left(1 + \lambda^2 + \left\|\tfrac{\mathrm{d}}{\mathrm{d}t}\alpha\right\|_{A(\mathbb{T};\mathbb{R})} + \lambda^\rho\right)\right) \leq 1.$$

This condition is satisfied for all $\lambda \leq \lambda_0$ if we choose λ_0 and κ sufficiently small. Similarly, for v_1, $v_2 \in M_\delta$ we have

$$\|\mathcal{S}_{\lambda,\omega}(v_1) - \mathcal{S}_{\lambda,\omega}(v_2)\|_{A^q} \leq C_{40}\|\mathcal{N}(v_1) - \mathcal{N}(v_2)\|_{A^q}$$
$$\leq C_{40}C_{44}\left(\varepsilon\lambda^{-1} + \lambda + \omega + \varepsilon + 2\lambda^{-(3q-3)/q}\delta\right)\|v_1 - v_2\|_{\mathcal{X}^q_{\lambda,\omega}}.$$

With the same choices of parameters as above, this yields that $\mathcal{S}_{\lambda,\omega}$ is a contraction on the set M_δ if

$$C_{40}C_{44}\left(2\lambda + \kappa\lambda^\rho + \lambda^2 + 2\lambda^{\rho-(3q-3)/q}\right) \leq \frac{1}{2},$$

which holds if $\lambda \leq \lambda_0$ and λ_0 is sufficiently small. In total, we thus obtain that, under the above choice of parameters, the solution map $\mathcal{S}_{\lambda,\omega}$ is a contractive self-mapping. Finally, the contraction mapping principle yields the existence of a fixed point $v \in \mathcal{X}^q_{\lambda,\omega}$ of $\mathcal{S}_{\lambda,\omega}$, and hence of a solution (v, p) to (4.49). Consequently, $(u, \mathfrak{p}) := (v + U, p)$ is a solution to (4.1) in the asserted function space. $\qquad\square$

Remark 4.3.4. The lower bound $\frac{\lambda^2}{\theta} \leq \omega$ on the angular velocity in (4.53) may seem strange since, from a physical point of view, the limit $\omega \to 0$, which corresponds to the case of a non-rotating body, seems uncritical. The lower bound on ω in the condition (4.53) is an artifact of the change of coordinates into the rotating frame of reference employed in the mathematical analysis of the linear problem, which leads to *a priori* estimates with constants that exhibit a singular behavior as $\omega \to 0$. As a consequence, a lower bound on ω is necessary in Theorem 4.3.3 to obtain existence of a solution via a fixed-point iteration. A similar observation was made in the investigation of a steady flow past a rotating and translating body carried out in [30]. From a mathematical point of view, it is therefore not surprising that the same effect appears in the more general time-periodic case investigated here.

5 Time-Periodic Fundamental Solutions

A classical concept in the theory of partial differential equations is the notion of fundamental solutions. Fundamental solutions to both the steady-state and the initial-value linearized Navier–Stokes equations have been studied for many decades, initiated by LORENTZ [80] and OSEEN [85]. In contrast, a fundamental solution to the time-periodic Stokes system was introduced just recently by KYED [76]. In this chapter, we extend this latter result in various directions. Firstly, we derive a time-periodic fundamental solution to the general linearized Navier–Stokes equations in both the Stokes and the Oseen case in dimension $n \geq 2$. The corresponding results were published in [21]. Secondly, we introduce a time-periodic fundamental solution for the vorticity field associated with a solution to the linearized Navier–Stokes equations in three dimensions.

These time-periodic fundamental solutions always consist of two parts: A *steady-state* part that coincides with the fundamental solution to the steady-state problem, and a second *purely periodic* part. While integrability properties and pointwise estimates of the former are well known from the theory of steady-state problems, we establish corresponding results for the purely periodic parts. These properties enable us to identify the fundamental solutions as regular distributions on a Schwartz–Bruhat space and express convolutions with the fundamental solutions in terms of classical integrals. Moreover, they facilitate the investigation of spatial decay of time-periodic solutions to both the linearized and the nonlinear Navier–Stokes equations. While the study of the first is straightforward, the latter is more involved and will be the topic of Chapter 6.

In Section 5.1 we recall the fundamental solutions to the steady-state Stokes and Oseen equations and we examine a fundamental solution to a Helmholtz equation with a drift term. Based on these steady-state fundamental solutions, we introduce time-periodic fundamental solutions to the Stokes and Oseen equations in Section 5.2 and establish pointwise estimates and integrability properties. In Section 5.3 we introduce and investigate a fundamental solution for the vorticity field in the three-dimensional case.

5.1 Classical Fundamental Solutions

In this section, we consider several steady-state problems. After recalling the fundamental solutions to the Laplace equation as well as to the steady-state Stokes and Oseen equations, we introduce a fundamental solution to a Helmholtz equation with a drift term. We subsequently derive estimates of the convolution of this fundamental solution with the Laplace fundamental solution, which is a preparation for the study of the time-periodic fundamental solutions in Section 5.2. At the end of this section, we collect specific results in the three-dimensional case.

5.1.1 The Stokes and Oseen Equations

Here we recall fundamental solutions to several time-independent problems and the basic concept of fundamental solutions. To begin with, we consider the Laplace equation

$$-\Delta u = f \quad \text{in } \mathbb{R}^n \tag{5.1}$$

for $n \geq 2$. A fundamental solution to this equation is the well-known *Laplace fundamental solution* given by

$$\Gamma_L : \mathbb{R}^n \setminus \{0\} \to \mathbb{R}, \quad \Gamma_L(x) := \begin{cases} -\dfrac{1}{2\pi} \log|x| & \text{if } n = 2, \\ \dfrac{1}{(n-2)\omega_n} |x|^{2-n} & \text{if } n > 2, \end{cases} \tag{5.2}$$

where ω_n denotes the surface area of the $(n-1)$-dimensional unit sphere in $\mathbb{R}^n$. Then $\Gamma_L \in \mathscr{S}'(\mathbb{R}^n)$ is a solution to $-\Delta\Gamma_L = \delta_{\mathbb{R}^n}$ in the sense of distributions. Due to this property, for $f \in \mathscr{S}(\mathbb{R}^n)$ a solution u to (5.1) can be computed explicitly by the convolution

$$u = \Gamma_L * f. \tag{5.3}$$

Since Γ_L is not only an abstract distribution in $\mathscr{S}'(\mathbb{R}^n)$ but represented by the function (5.2), this convolution is a classical integral. This fact can be exploited to derive properties of the solution u from properties of the fundamental solution and the right-hand side f, which can be seen as the main advantage of this kind of representation formula.

Of course, this approach was also applied in the theory of Navier–Stokes equations. We consider the stationary linearized Navier–Stokes equations

$$\begin{cases} -\Delta v - \lambda\partial_1 v + \nabla p = f & \text{in } \mathbb{R}^n, \\ \operatorname{div} v = 0 & \text{in } \mathbb{R}^n \end{cases} \tag{5.4}$$

for a parameter $\lambda \in \mathbb{R}$. For $\lambda = 0$ this system is called the steady-state Stokes system, and for $\lambda \neq 0$ it is called the steady-state Oseen system. Here, $f : \mathbb{R}^n \to \mathbb{R}^n$ is a given right-hand side, and the solution consists of a velocity field $v : \mathbb{R}^n \to \mathbb{R}^n$ and a pressure field $p : \mathbb{R}^n \to \mathbb{R}$. By analogy to (5.3), a fundamental solution Φ_0 to (5.4) is a distribution such that we can express the solution (v, p) to (5.4) as

$$\begin{pmatrix} v \\ p \end{pmatrix} := \Phi_0 * f. \tag{5.5}$$

Therefore, the fundamental solution Φ_0 is a tensor field

$$\Phi_0 := \begin{pmatrix} \Gamma_{0,11}^\lambda & \cdots & \Gamma_{0,1n}^\lambda \\ \vdots & \ddots & \vdots \\ \Gamma_{0,n1}^\lambda & \cdots & \Gamma_{0,nn}^\lambda \\ \gamma_{0,1} & \cdots & \gamma_{0,n} \end{pmatrix} \in \mathscr{S}'(\mathbb{R}^n)^{(n+1)\times n}. \tag{5.6}$$

Then, the fundamental solution consists of the *velocity fundamental solution* $\Gamma_0^\lambda = (\Gamma_{0,j\ell}^\lambda)_{j,\ell=1}^n$ and the *pressure fundamental solution* $\gamma_0 = (\gamma_{0,j})_{j=1}^n$, which leads to the representation formulas

$$v = \Gamma_0^\lambda * f, \qquad p = \gamma_0 * f.$$

The components of Φ_0 satisfy the distributional equations

$$\begin{cases} -\Delta\Gamma_{0,j\ell}^\lambda - \lambda\partial_1\Gamma_{0,j\ell}^\lambda + \partial_j\gamma_{0,\ell} = \delta_{j\ell}\delta_{\mathbb{R}^n}, \\ \partial_h\Gamma_{0,h\ell}^\lambda = 0 \end{cases} \tag{5.7}$$

for all $j, \ell = 1, \ldots, n$, where we employ the Einstein summation convention in the second line. The fundamental solution to this problem is well known, but it heavily depends on whether $\lambda = 0$ or $\lambda \neq 0$. In the Stokes case ($\lambda = 0$), a velocity fundamental solution $\Gamma_0^\lambda = \Gamma^{\mathrm{S}} \colon \mathbb{R}^n \setminus \{0\} \to \mathbb{R}^{n\times n}$ to (5.7) is given by

$$\Gamma_{j\ell}^{\mathrm{S}}(x) := \begin{cases} \dfrac{1}{2\omega_n}\left(-\delta_{j\ell}\log|x| + \dfrac{x_j x_\ell}{|x|^2}\right) & \text{if } n = 2, \\[3mm] \dfrac{1}{2\omega_n}\left(\delta_{j\ell}\dfrac{1}{n-2}|x|^{2-n} + \dfrac{x_j x_\ell}{|x|^n}\right) & \text{if } n \geq 3; \end{cases} \tag{5.8}$$

see [42, Section IV.2] for example. In the Oseen case ($\lambda \neq 0$), a velocity fundamental solution $\Gamma_0^\lambda = \Gamma^{\mathrm{O}}$ to (5.7) is given by

$$\Gamma^{\mathrm{O}} \colon \mathbb{R}^n \setminus \{0\} \to \mathbb{R}^{n\times n}, \qquad \Gamma_{j\ell}^{\mathrm{O}}(x) := \left[\delta_{j\ell}\Delta - \partial_j\partial_\ell\right]\Psi_0^\lambda(x) \tag{5.9}$$

with

$$\Psi_0^\lambda(x) = \frac{1}{\lambda}\int_0^{x_1}\left[\Gamma_{\mathrm{L}}(\tau, x_2, \ldots, x_n) - \Xi(\tau, x_2, \ldots, x_n)\right]\mathrm{d}\tau$$

$$+ \frac{1}{4\pi}\int_0^{-x_2}(\tau + x_2)K_0(\lambda|\tau|)\,\mathrm{d}\tau$$

if $n = 2$, and

$$\Psi_0^\lambda(x) = \frac{1}{\lambda}\int_{-\infty}^{x_1}\left[\Gamma_{\mathrm{L}}(\tau, x_2, \ldots, x_n) - \Xi(\tau, x_2, \ldots, x_n)\right]\mathrm{d}\tau$$

if $n \geq 3$, where Γ_{L} is the Laplace fundamental solution defined in (5.2) and

$$\Xi(x) := \frac{1}{2\pi}\left(\frac{\lambda}{4\pi|x|}\right)^{\frac{n-2}{2}}K_{\frac{n-2}{2}}\left(\frac{\lambda}{2}|x|\right)e^{-\frac{\lambda}{2}x_1};$$

see [42, Section VII.3] for example. Here, K_ν denotes the modified Bessel function of the second kind; see Section A.1.1. In both the Stokes and the Oseen case, a pressure fundamental solution is given by

$$\gamma_0 : \mathbb{R}^n \smallsetminus \{0\} \to \mathbb{R}^n, \quad \gamma_{0,j}(x) := -\frac{1}{\omega_n} \frac{x_j}{|x|^n} = \partial_j \Gamma_{\mathrm{L}}(x). \tag{5.10}$$

Finally, Γ_0^λ and γ_0 constitute a fundamental solution in the form (5.6) to the steady-state linearized Navier–Stokes equations (5.4).

5.1.2 The Helmholtz Equation with a Drift Term

Next we study a fundamental solution to the equation

$$i\eta v - \Delta v - \lambda \partial_1 v = f \qquad \text{in } \mathbb{R}^n \tag{5.11}$$

for given parameters $\lambda \in \mathbb{R}$ and $\eta \in \mathbb{R} \smallsetminus \{0\}$. Note that for $\lambda = 0$ this is the classical Helmholtz equation. First of all, let us treat this special case and consider the function

$$\Gamma_{\mathrm{H}}^\mu : \mathbb{R}^n \smallsetminus \{0\} \to \mathbb{C}, \quad \Gamma_{\mathrm{H}}^\mu(x) := \frac{i}{4} \left(\frac{\sqrt{-\mu}}{2\pi|x|} \right)^{\frac{n-2}{2}} H_{\frac{n-2}{2}}^{(1)} \left(\sqrt{-\mu}\, |x| \right)$$

for $\mu \in \mathbb{C} \smallsetminus \mathbb{R}$. Here $H_\nu^{(1)}$ denotes the Hankel function of the first kind (see Subsection A.1.1) and $\sqrt{z}$ is the square root of z with *nonnegative* imaginary part. Then Γ_{H}^μ is a fundamental solution to the Helmholtz equation

$$\mu \Gamma_{\mathrm{H}}^\mu - \Delta \Gamma_{\mathrm{H}}^\mu = \delta_{\mathbb{R}^n};$$

see [98, Chapter 5.8] for example. We shall use this fact to find a fundamental solution to (5.11). For $\lambda \in \mathbb{R}$ and $\eta \neq 0$, we define the function $\Gamma_{\mathrm{H}}^{\eta,\lambda} : \mathbb{R}^n \smallsetminus \{0\} \to \mathbb{C}$ by

$$\Gamma_{\mathrm{H}}^{\eta,\lambda}(x) := \Gamma_{\mathrm{H}}^\mu(x)\, e^{-\frac{\lambda}{2}x_1} = \frac{i}{4} \left(\frac{\sqrt{-\mu}}{2\pi|x|} \right)^{\frac{n-2}{2}} H_{\frac{n-2}{2}}^{(1)} \left(\sqrt{-\mu}\, |x| \right) e^{-\frac{\lambda}{2}x_1} \tag{5.12}$$

with $\mu := \mu(\eta, \lambda) := (\lambda/2)^2 + i\eta \in \mathbb{C} \smallsetminus \mathbb{R}$. In order to see that $\Gamma_{\mathrm{H}}^{\eta,\lambda}$ is a fundamental solution to (5.11), we further analyze the term $\sqrt{-\mu}$ in the following lemma. Its proof mainly relies on an explicit computation of this quantity.

Lemma 5.1.1. *Let $\eta_0 > 0$ and $\lambda \in \mathbb{R}$. Then there exists a constant $C_{45} = C_{45}(\lambda, \eta_0) > 0$ such that*

$$\mathrm{Im}(\sqrt{-\mu}) - \frac{|\lambda|}{2} \geq C_{45}|\eta|^{\frac{1}{2}} \tag{5.13}$$

for all $\eta \in \mathbb{R}$ with $|\eta| \geq \eta_0$ and $\mu = (\lambda/2)^2 + i\eta$.

Proof. For $\lambda = 0$ the statement follows directly with $C_{45} = 1/\sqrt{2}$. If we assume $\lambda \neq 0$, then

$$\sqrt{-\mu} = \left((\lambda/2)^4 + \eta^2\right)^{\frac{1}{4}} \exp\left(\frac{i}{2}\left(\pi + \arctan(4\eta\lambda^{-2})\right)\right).$$

Using the identity $2\cos^2(x) = 1 + \cos(2x)$, we thus obtain

$$\mathrm{Im}(\sqrt{-\mu}) = \left((\lambda/2)^4 + \eta^2\right)^{\frac{1}{4}} \sin\left(\frac{1}{2}\left(\pi + \arctan(4\eta\lambda^{-2})\right)\right)$$

$$= \left((\lambda/2)^4 + \eta^2\right)^{\frac{1}{4}} \cos\left(\frac{1}{2}\arctan(4\eta\lambda^{-2})\right)$$

$$= \left((\lambda/2)^4 + \eta^2\right)^{\frac{1}{4}} \frac{1}{\sqrt{2}}\left(1 + \cos\left(\arctan(4\eta\lambda^{-2})\right)\right)^{\frac{1}{2}}.$$

We employ the identity $\cos(\arctan(x)) = (1 + x^2)^{-1/2}$ to further deduce

$$\mathrm{Im}(\sqrt{-\mu}) = \left((\lambda/2)^4 + \eta^2\right)^{\frac{1}{4}} \frac{1}{\sqrt{2}}\left(1 + \left(1 + (4\eta\lambda^{-2})^2\right)^{-\frac{1}{2}}\right)^{\frac{1}{2}}$$

$$= \frac{|\lambda|}{2} \frac{1}{\sqrt{2}}\left(\left(1 + \frac{\eta^2}{(\lambda/2)^4}\right)^{\frac{1}{2}} + 1\right)^{\frac{1}{2}}.$$

Consequently, we have $\mathrm{Im}(\sqrt{-\mu}) - |\lambda|/2 > 0$ for $\eta \neq 0$, and

$$\lim_{|\eta| \to \infty} \frac{\mathrm{Im}(\sqrt{-\mu}) - |\lambda|/2}{|\eta|^{\frac{1}{2}}} = \frac{1}{\sqrt{2}}.$$

This implies the assertion since $|\eta| \geq \eta_0$. $\qquad\square$

With the help of the previous lemma and the results on Hankel functions from Subsection A.1.1 and the estimates from Subsection A.1.2, we can now establish the following pointwise estimates of $\Gamma_{\mathrm{H}}^{\eta,\lambda}$ and its derivatives.

Lemma 5.1.2. *For all ε, $\eta_0 > 0$, $\lambda \in \mathbb{R}$ and $m \in \mathbb{N}_0$, there exists a constant $C_{46} = C_{46}(n, \varepsilon, \lambda, \eta_0) > 0$ such that*

$$\left| \Gamma_{\mathrm{H}}^{\eta,\lambda}(x) \right| \le C_{46} |\mu|^{\frac{n-3}{4}} |x|^{\frac{1-n}{2}} \, \mathrm{e}^{-C_{45}|\eta|^{\frac{1}{2}}|x|}, \tag{5.14}$$

$$\left| \nabla^m \Gamma_{\mathrm{H}}^{\eta,\lambda}(x) \right| \le C_{46} |\eta|^{-1} |x|^{-n-m} \, \mathrm{e}^{-C_{45}|\eta|^{\frac{1}{2}}|x|/2} \tag{5.15}$$

for all $x \in \mathbb{R}^n$ with $|x| \ge \varepsilon$ and all $\eta \in \mathbb{R}$ with $|\eta| \ge \eta_0$. Here, C_{45} is the constant from Lemma 5.1.1.

Proof. Recalling the definition of $\Gamma_{\mathrm{H}}^{\eta,\lambda}$ in (5.12) and estimate (A.67) for the Hankel functions, we obtain

$$\left| \Gamma_{\mathrm{H}}^{\eta,\lambda}(x) \right| \le c_0 |\mu|^{\frac{n-2}{4}} |x|^{\frac{2-n}{2}} \left| H_{\frac{n-2}{2}}^{(1)} \left(\sqrt{-\mu} \cdot |x| \right) \right| \mathrm{e}^{\frac{|\lambda|}{2}|x|}$$

$$\le c_1 |\mu|^{\frac{n-3}{4}} |x|^{\frac{1-n}{2}} \, \mathrm{e}^{-\mathrm{Im}(\sqrt{-\mu})|x| + \frac{|\lambda|}{2}|x|}.$$

Estimating now the exponential function with the help of (5.13), we immediately arrive at (5.14). Another elementary estimate implies

$$\left| \Gamma_{\mathrm{H}}^{\eta,\lambda}(x) \right| \le c_1 |\eta|^{-1} |x|^{-n} \left(|\eta|^{\frac{1}{2}}|x| \right)^{\frac{n+1}{2}} \mathrm{e}^{-C_{45}|\eta|^{\frac{1}{2}}|x|} \le c_2 |\eta|^{-1} |x|^{-n} \, \mathrm{e}^{-C_{45}|\eta|^{\frac{1}{2}}|x|/2},$$

which is (5.15) for $m = 0$. To derive (5.15) for general $m \in \mathbb{N}_0$, we compute the derivatives of $\Gamma_{\mathrm{H}}^{\eta,\lambda}$. By the Leibniz rule, for $\alpha \in \mathbb{N}_0^n$ we have

$$D^\alpha \Gamma_{\mathrm{H}}^{\eta,\lambda}(x) = \frac{i}{4} \left(\frac{\sqrt{-\mu}}{2\pi} \right)^{\frac{n-2}{2}} \sum_{\beta \le \alpha} c_\beta D^\beta \left[|x|^{\frac{2-n}{2}} \mathrm{e}^{-\frac{\lambda}{2}x_1} \right] D^{\alpha-\beta} \left[H_{\frac{n-2}{2}}^{(1)} \left(\sqrt{-\mu} |x| \right) \right].$$

Estimates of the derivatives appearing on the right-hand side can be found in Lemma A.1.5 and Lemma A.1.3. We thus deduce

$$\left| D^\alpha \Gamma_{\mathrm{H}}^{\eta,\lambda}(x) \right| \le c_3 |\mu|^{\frac{n-2}{4}} \sum_{\beta \le \alpha} |x|^{\frac{2-n}{2}} \mathrm{e}^{-\frac{\lambda}{2}x_1} |\mu|^{\frac{2|\alpha-\beta|-1}{4}} |x|^{-\frac{1}{2}} \mathrm{e}^{-\mathrm{Im}(\sqrt{-\mu})|x|}$$

$$\le c_4 |\mu|^{\frac{n+2|\alpha|-3}{4}} |x|^{\frac{1-n}{2}} \, \mathrm{e}^{-\mathrm{Im}(\sqrt{-\mu})|x| + \frac{|\lambda|}{2}|x|},$$

where we used $|\mu| \ge |\eta| \ge \eta_0$. By employing Lemma 5.1.1 and the inequality $|\mu| \le c_5 |\eta|$ with $c_5 = c_5(\lambda, \eta_0) > 0$, we arrive at

$$\left| D^\alpha \Gamma_{\mathrm{H}}^{\eta,\lambda}(x) \right| \le c_6 |\eta|^{-1} |x|^{-(n+|\alpha|)} \left(|\eta|^{\frac{1}{2}}|x| \right)^{\frac{n+2|\alpha|+1}{2}} \mathrm{e}^{-C_{45}|\eta|^{\frac{1}{2}}|x|}$$

$$\le c_7 |\eta|^{-1} |x|^{-(n+|\alpha|)} \, \mathrm{e}^{-C_{45}|\eta|^{\frac{1}{2}}|x|/2},$$

which implies (5.15). This completes the proof. $\qquad\qquad\square$

Next we derive an estimate that describes the asymptotic behavior of $\Gamma_{\mathrm{H}}^{\eta,\lambda}(x)$ as $|x| \to 0$. It will be a direct consequence of the according estimates for Hankel functions given in Lemma A.1.1.

Lemma 5.1.3. *For all $\lambda, \eta \in \mathbb{R}$ with $\eta \neq 0$ and $R \in (0,1)$ there is a constant $C_{47} = C_{47}(n, R, \lambda) > 0$ such that*

$$|\Gamma_{\mathrm{H}}^{\eta,\lambda}(x)| \leq C_{47}|x|^{-n+2}\, e^{\frac{\lambda}{2}x_1} \qquad\qquad \text{if } n > 2, \qquad\qquad (5.16)$$

$$|\Gamma_{\mathrm{H}}^{\eta,\lambda}(x)| \leq C_{47}|\log(|\mu|^{\frac{1}{2}}|x|)|\, e^{\frac{\lambda}{2}x_1} \qquad\qquad \text{if } n = 2, \qquad\qquad (5.17)$$

for all $x \in \mathbb{R}^n$ with $|\mu|^{\frac{1}{2}}|x| \leq R$.

Proof. From the definition of $\Gamma_{\mathrm{H}}^{\eta,\lambda}$ in (5.12) we directly conclude

$$|\Gamma_{\mathrm{H}}^{\eta,\lambda}(x)| \leq c_0|\mu|^{\frac{n-2}{4}}|x|^{-\frac{n+2}{2}}\left|H_{\frac{n-2}{2}}^{(1)}\left(\sqrt{-\mu}\,|x|\right)\right| e^{-\frac{\lambda}{2}x_1}.$$

Now (5.16) and (5.17) are direct consequences of the estimates (A.68) and (A.69) for Hankel functions. $\qquad\qquad\square$

Now we can verify that $\Gamma_{\mathrm{H}}^{\eta,\lambda}$ is an element of $\mathscr{S}'(\mathbb{R}^n)$ and a fundamental solution to (5.11), that is, it satisfies the equation

$$i\eta\, \Gamma_{\mathrm{H}}^{\eta,\lambda} - \Delta\Gamma_{\mathrm{H}}^{\eta,\lambda} - \lambda\partial_1\Gamma_{\mathrm{H}}^{\eta,\lambda} = \delta_{\mathbb{R}^n}. \qquad\qquad (5.18)$$

The proof will mainly be based on the estimates of $\Gamma_{\mathrm{H}}^{\eta,\lambda}$ derived in the previous lemmas.

Lemma 5.1.4. *For all $\lambda \in \mathbb{R}$ and $\eta \in \mathbb{R} \setminus \{0\}$, the function $\Gamma_{\mathrm{H}}^{\eta,\lambda}$ is a fundamental solution in $\mathscr{S}'(\mathbb{R}^n)$ to (5.18), it is an element of $\mathrm{L}^1(\mathbb{R}^n)$ and satisfies*

$$\Gamma_{\mathrm{H}}^{\eta,\lambda} = \mathscr{F}_{\mathbb{R}^n}^{-1}\left[\frac{1}{i\eta + |\xi|^2 - i\lambda\xi_1}\right]. \qquad\qquad (5.19)$$

Proof. From Lemma 5.1.2 and Lemma 5.1.3, we immediately conclude $\Gamma_{\mathrm{H}}^{\eta,\lambda} \in \mathrm{L}^1(\mathbb{R}^n) \subseteq \mathscr{S}'(\mathbb{R}^n)$. Now, a short computation leads to

$$\left[i\eta - \Delta - \lambda\partial_1\right]\Gamma_{\mathrm{H}}^{\eta,\lambda}$$

$$= \left(i\eta\Gamma_{\mathrm{H}}^{\mu} - \Delta\Gamma_{\mathrm{H}}^{\mu} + \lambda\partial_1\Gamma_{\mathrm{H}}^{\mu} - \frac{\lambda^2}{4}\Gamma_{\mathrm{H}}^{\mu} - \lambda\partial_1\Gamma_{\mathrm{H}}^{\mu} + \frac{\lambda^2}{2}\Gamma_{\mathrm{H}}^{\mu}\right)e^{-\frac{\lambda}{2}x_1}$$

$$= \left(-\Delta\Gamma_{\mathrm{H}}^{\mu} + \mu\Gamma_{\mathrm{H}}^{\mu}\right)e^{-\frac{\lambda}{2}x_1} = \delta_{\mathbb{R}^n}.$$

Therefore, $\Gamma_{\mathrm{H}}^{\eta,\lambda}$ is a fundamental solution to (5.18). Applying the Fourier transform to this identity, we obtain $(i\eta + |\xi|^2 - i\lambda\xi_1)\mathscr{F}_{\mathbb{R}^n}[\Gamma_{\mathrm{H}}^{\eta,\lambda}] = 1$. Due to $\eta \neq 0$, this implies (5.19) and completes the proof. $\qquad\qquad\square$

5.1.3 A Convolution Estimate

Here we consider the convolution of $\Gamma_{\mathrm{H}}^{\eta,\lambda}$ with the Laplace fundamental solution Γ_{L} defined in (5.2). The presented result will be the key to derive pointwise estimates of the time-periodic fundamental solution later. The proof is based on the estimates of $\Gamma_{\mathrm{H}}^{\eta,\lambda}$ from Lemma 5.1.2 and Lemma 5.1.3. Moreover, for $m \geq 1$ we frequently use the estimate

$$|\nabla^m \Gamma_{\mathrm{L}}(x)| \leq C_{48}|x|^{2-n-m}, \tag{5.20}$$

which is a direct consequence of (5.2).

Lemma 5.1.5. *Let $\lambda \in \mathbb{R}$ and $\eta \in \mathbb{R} \setminus \{0\}$. Then the convolution integral*

$$[\Gamma_{\mathrm{L}} * \Gamma_{\mathrm{H}}^{\eta,\lambda}](x) = \int_{\mathbb{R}^n} \Gamma_{\mathrm{L}}(x - y) \, \Gamma_{\mathrm{H}}^{\eta,\lambda}(y) \, \mathrm{d}y \tag{5.21}$$

exists for all $x \in \mathbb{R}^n \setminus \{0\}$ and $\Gamma_{\mathrm{L}} * \Gamma_{\mathrm{H}}^{\eta,\lambda} \in \mathrm{L}_{\mathrm{loc}}^1(\mathbb{R}^n)$. Moreover, $\Gamma_{\mathrm{L}} * \Gamma_{\mathrm{H}}^{\eta,\lambda} \in \mathrm{C}^\infty(\mathbb{R}^n \setminus \{0\}) \cap \mathrm{W}_{\mathrm{loc}}^{k,1}(\mathbb{R}^n)$ for all $k \in \mathbb{N}_0$, and for all $\beta \in \mathbb{N}_0^n$ with $|\beta| \geq 1$ and all $\varepsilon, \eta_0 > 0$ there exists a constant $C_{49} = C_{49}(n, \lambda, \eta_0, \beta, \varepsilon) > 0$ such that

$$\left| D^\beta [\Gamma_{\mathrm{L}} * \Gamma_{\mathrm{H}}^{\eta,\lambda}](x) \right| \leq C_{49} |\eta|^{-1} |x|^{2-n-|\beta|} \tag{5.22}$$

for all $\eta \in \mathbb{R}$ with $|\eta| \geq \eta_0$ and all $x \in \mathbb{R}^n$ with $|x| \geq \varepsilon$.

Proof. In virtue of Lemma A.2.1, the estimates from Lemma 5.1.2 and Lemma 5.1.3 together with (5.20) imply that the right-hand side of (5.21) is well defined for $x \neq 0$ and that $\Gamma_{\mathrm{L}} * \Gamma_{\mathrm{H}}^{\eta,\lambda} \in \mathrm{L}_{\mathrm{loc}}^1(\mathbb{R}^n) \cap \mathrm{C}^1(\mathbb{R}^n \setminus \{0\})$ with

$$\partial_j [\Gamma_{\mathrm{L}} * \Gamma_{\mathrm{H}}^{\eta,\lambda}](x) = \int_{\mathbb{R}^n} \partial_j \Gamma_{\mathrm{L}}(x - y) \, \Gamma_{\mathrm{H}}^{\eta,\lambda}(y) \, \mathrm{d}y. \tag{5.23}$$

Now fix $\varepsilon > 0$ and consider some $x \in \mathbb{R}^n$ with $|x| \geq \varepsilon$. Put $R := \frac{|x|}{2}$. Let $\chi \in \mathrm{C}_0^\infty(\mathbb{R}; \mathbb{R})$ be a cut-off function with

$$\chi(r) = 1 \quad \text{if } 1 \leq |r| \leq 3, \qquad \chi(r) = 0 \quad \text{if } 0 \leq |r| \leq \frac{1}{2} \text{ or } |r| \geq 4,$$

and define $\chi_R : \mathbb{R}^n \to \mathbb{R}$ by $\chi_R(y) := \chi(R^{-1}|y|)$. We decompose (5.23) as

$$\partial_j [\Gamma_{\mathrm{L}} * \Gamma_{\mathrm{H}}^{\eta,\lambda}](x) = I_1(x) + I_2(x) + I_3(x)$$

where

$$I_1(x) = \int_{B_R} \partial_j \Gamma_{\mathrm{L}}(x-y)\, \Gamma_{\mathrm{H}}^{\eta,\lambda}(y) \left(1 - \chi_R(y)\right) dy,$$

$$I_2(x) = \int_{B^{3R}} \partial_j \Gamma_{\mathrm{L}}(x-y)\, \Gamma_{\mathrm{H}}^{\eta,\lambda}(y) \left(1 - \chi_R(y)\right) dy,$$

$$I_3(x) = \int_{B_{4R} \setminus B_{R/2}} \partial_j \Gamma_{\mathrm{L}}(x-y)\, \Gamma_{\mathrm{H}}^{\eta,\lambda}(y)\, \chi_R(y)\, dy.$$

We consider each term separately. In virtue of the decay properties of Γ_{L} from (5.20) and the estimates of $\Gamma_{\mathrm{H}}^{\eta,\lambda}$ in Lemma 5.1.3 and Lemma 5.1.2, we see that $I_1 \in C^\infty(\mathbb{R}^n \setminus \{0\}) \cap W_{\mathrm{loc}}^{k,1}(\mathbb{R}^n)$ for all $k \in \mathbb{N}_0$ by Lemma A.2.1. Since $|y| \le R$ implies $|x - y| \ge |x| - |y| \ge R/2$, estimate (5.20) yields

$$\left| D^\alpha I_1(x) \right| = \left| \int_{B_R} \partial_j D^\alpha \Gamma_{\mathrm{L}}(x-y)\, \Gamma_{\mathrm{H}}^{\eta,\lambda}(y) \left(1 - \chi_R(y)\right) dy \right|$$

$$\le c_0 \int_{B_R} \left| \partial_j D^\alpha \Gamma_{\mathrm{L}}(x-y) \right| \left| \Gamma_{\mathrm{H}}^{\eta,\lambda}(y) \right| dy$$

$$\le c_1 \int_{\mathbb{R}^n} R^{1-n-|\alpha|} \left| \Gamma_{\mathrm{H}}^{\eta,\lambda}(y) \right| dy.$$

We split this integral at a radius $\delta = \frac{1}{2}|\mu|^{-\frac{1}{2}}$. With (5.14) we obtain

$$\int_{B^\delta} \left| \Gamma_{\mathrm{H}}^{\eta,\lambda}(y) \right| dy \le c_2 |\mu|^{\frac{n-3}{4}} \int_{B^\delta} |y|^{\frac{1-n}{2}} e^{-C_{45}|\eta|^{\frac{1}{2}}|y|}\, dy$$

$$\le c_3 |\mu|^{-1} \int_{B^{1/2}} |y|^{\frac{1-n}{2}} e^{-C_{45}|\eta|^{\frac{1}{2}}|\mu|^{-\frac{1}{2}}|y|}\, dy \le c_4 |\mu|^{-1} \int_{B^{1/2}} |y|^{\frac{1-n}{2}}\, dy \le c_5 |\eta|^{-1}.$$

Similarly, from Lemma 5.1.3 we deduce

$$\int_{B_\delta} \left| \Gamma_{\mathrm{H}}^{\eta,\lambda}(y) \right| dy \le c_6 \int_{B_\delta} |y|^{-n+2} e^{\frac{\lambda}{2}y_1}\, dy$$

$$\le c_7 |\mu|^{-1} \int_{B_{1/2}} |y|^{-n+2} e^{\frac{\lambda}{2}|\mu|^{-\frac{1}{2}}y_1}\, dy \le c_8 |\eta|^{-1}$$

in the case $n > 2$, and

$$\int_{B_\delta} \left| \Gamma_{\mathrm{H}}^{\eta,\lambda}(y) \right| dy \le c_9 \int_{B_\delta} \left| \log(|\mu|^{\frac{1}{2}}|y|) \right| e^{\frac{\lambda}{2}y_1}\, dy$$

$$\le c_{10} |\mu|^{-1} \int_{B_{1/2}} \left| \log(|y|) \right| e^{\frac{\lambda}{2}|\mu|^{-\frac{1}{2}}y_1}\, dy \le c_{11} |\eta|^{-1}.$$

in the case $n = 2$, where we used $|\lambda|/2 \le |\mu|^{\frac{1}{2}}$ in both cases. Collecting these estimates, we conclude

$$\left| D^\alpha I_1(x) \right| \le c_{12} |\eta|^{-1} R^{1-n-|\alpha|}.$$

Now let us turn to I_2. As above, Lemma A.2.1 implies $I_2 \in C^\infty(\mathbb{R}^n \setminus \{0\}) \cap W^{k,1}_{\text{loc}}(\mathbb{R}^n)$ for all $k \in \mathbb{N}_0$. In order to estimate $D^\alpha I_2$, we utilize (5.20) and Lemma 5.1.2 again. Since $|y| \ge 3R \ge 3\varepsilon/2$ implies $|\mu|^{\frac{1}{2}}|y| \ge 3\eta_0\varepsilon/2$, we can employ estimate (5.15) with $m = 0$, which directly yields

$$\left| D^\alpha I_2(x) \right| \le c_{13} \int_{B^{3R}} |\partial_j D^\alpha \Gamma_{\text{L}}(x-y)| \, |\Gamma_{\text{H}}^{\eta,\lambda}(y)| \, dy$$

$$\le c_{14} \int_{B^{3R}} |x-y|^{1-n-|\alpha|} |\eta|^{-1} |y|^{-n} \, dy$$

$$\le c_{15} |\eta|^{-1} R^{1-n-|\alpha|}.$$

Furthermore, from the decay estimate of $\Gamma_{\text{H}}^{\eta,\lambda}$ from (5.15) we conclude $I_3 \in C^\infty(\mathbb{R}^n \setminus \{0\}) \cap W^{k,1}_{\text{loc}}(\mathbb{R}^n)$ for all $k \in \mathbb{N}_0$ by Lemma A.2.1, and we have

$$D^\alpha I_3(x) = \int_{B_{4R} \setminus B_{R/2}} \partial_j \Gamma_{\text{L}}(x-y) \, D^\alpha \big[\Gamma_{\text{H}}^{\eta,\lambda} \chi_R \big](y) \, dy.$$

Since $|y| \ge R/2 \ge \varepsilon/4$ implies $|\mu|^{\frac{1}{2}}|y| \ge \eta_0^{\frac{1}{2}}\varepsilon/4$, inequality (5.15) and the Leibniz rule lead to the estimate

$$\left| D^\alpha I_3(x) \right| \le c_{16} \int_{B_{4R} \setminus B_{R/2}} |x-y|^{1-n} \sum_{k=0}^{|\alpha|} |\eta|^{-1} |y|^{-n-k} R^{-(|\alpha|-k)} \, dy$$

$$\le c_{17} |\eta|^{-1} R^{-n-|\alpha|} \int_{B_{6R}(x)} |x-y|^{1-n} \, dy \le c_{18} |\eta|^{-1} R^{1-n-|\alpha|}.$$

In total, we have shown $\Gamma_{\text{L}} * \Gamma_{\text{H}}^{\eta,\lambda} \in C^\infty(\mathbb{R}^n \setminus \{0\}) \cap W^{k,1}_{\text{loc}}(\mathbb{R}^n)$ for all $k \in \mathbb{N}_0$, and since $|x| = 2R$, we finally conclude (5.22) with $\beta = \alpha + e_j$ by collecting the estimates for $D^\alpha I_1$, $D^\alpha I_2$ and $D^\alpha I_3$. $\qquad \square$

5.1.4 The Three-Dimensional Case

One of the main goals in Chapter 6 is to derive spatial decay estimates for the solution to the Navier–Stokes equations. These estimates mainly rely on a representation of the solution via the fundamental solution. For the derivation of pointwise information, we need pointwise properties of the fundamental solution. We restrict ourselves to the Oseen case $\lambda \neq 0$ in the following. Then the steady-state velocity fundamental solution is given by the three-dimensional Oseen fundamental solution $\Gamma_0^\lambda = \Gamma^O$. The function Ψ_0^λ in formula (5.9) can then be simplified since the associated modified Bessel function $K_{1/2}$ is given by

$$K_{1/2}(z) = \sqrt{\frac{\pi}{2z}}\, e^{-z}\,.$$

We thus obtain

$$\Gamma_{0,j\ell}^\lambda(x) = \frac{1}{4\pi\lambda}\big[\delta_{j\ell}\Delta - \partial_j\partial_\ell\big] \int_0^{s(\lambda x)/2} \frac{1 - e^{-\tau}}{\tau}\, d\tau, \tag{5.24}$$

where

$$s(x) := |x| + x_1.$$

One can derive the following estimates.

Theorem 5.1.6. *Let $n = 3$ and $\lambda \in \mathbb{R} \setminus \{0\}$. For all $m \in \mathbb{N}_0$ and $\varepsilon > 0$ there exists a constant $C_{50} > 0$ such that*

$$\forall |x| \geq \varepsilon: \quad |\nabla^m \Gamma_0^\lambda(x)| \leq C_{50}\big[|x|(1 + s(\lambda x))\big]^{-1-\frac{m}{2}}. \tag{5.25}$$

Proof. We refer to [26, Lemma 3.2]. $\qquad\square$

Observe that the appearance of the term $s(\lambda x)$ in (5.24) and (5.25) results in an anisotropic behavior of $\Gamma_0^\lambda(x)$ as $|x| \to \infty$. More precisely, if we consider $\lambda = 1$ for the moment, in all directions except of the negative x_1 axis, that is, on sets of the form

$$\big\{x \in \mathbb{R}^n \mid x_1 = \alpha|x|\big\} = \big\{x \in \mathbb{R}^n \mid s(x) = (1 + \alpha)|x|\big\}, \qquad \alpha \in (-1, 1],$$

the fundamental solution Γ_0^λ decays like $|x|^{-2}$. However, the above estimate merely yields decay of order $|x|^{-1}$ on parabola-like wake regions of the form

$$\big\{x \in \mathbb{R}^n \mid |x|^2 - x_1^2 + 2\beta x_1 \leq \beta^2\big\} = \big\{x \in \mathbb{R}^n \mid s(x) \leq \beta\big\}, \qquad \beta > 0.$$

As mentioned before, the strength of a fundamental solution is that it yields a solution formula via convolution. In order to exploit the full potential of the anisotropic decay properties of Γ_0^λ from Theorem 5.1.6, one thus has to control convolutions with functions satisfying similar estimates. The careful study of such kind of convolutions is an involved task, which was carried out by FARWIG [26, 27] in dimension $n = 3$, and later by KRAČMAR, NOVOTNÝ and POKORNÝ [71] in the general n-dimensional case. We collect some of their results in the following theorem, which gives estimates of convolutions with Γ_0^λ and $\nabla \Gamma_0^\lambda$.

Theorem 5.1.7. *Let $n = 3$, $A \in [2, \infty)$ and $B \in [0, \infty)$, and let $g \in \mathrm{L}^\infty(\mathbb{R}^3)$ such that $g(x) \le M(1 + |x|)^{-A}(1 + s(x))^{-B}$. Then there exists a constant $C_{51} = C_{51}(A, B, \lambda) > 0$ with the following properties:*

1. If $A + \min\{1, B\} > 3$, then

$$\left\| |\Gamma_0^\lambda| * g(x) \right\| \le C_{51} M \left[(1 + |x|)(1 + s(\lambda x)) \right]^{-1}.$$

2. If $A + \min\{1, B\} > 3$ and $A + B \ge 7/2$, then

$$\left\| |\nabla \Gamma_0^\lambda| * g(x) \right\| \le C_{51} M \left[(1 + |x|)(1 + s(\lambda x)) \right]^{-3/2}.$$

3. If $A + B < 3$, then

$$\left\| |\nabla \Gamma_0^\lambda| * g(x) \right\| \le C_{51} M (1 + |x|)^{-(A+B)/2} \left(1 + s(\lambda x)\right)^{-(A+B-1)/2}.$$

Proof. These are special cases of [71, Theorems 3.1 and 3.2]. □

One can also show the following integrability properties of Γ_0^λ.

Theorem 5.1.8. *Let $n = 3$ and $\lambda \in \mathbb{R} \setminus \{0\}$. Then*

$$\forall q \in [1, 3): \qquad \Gamma_0^\lambda \in \mathrm{L}_{\mathrm{loc}}^q(\mathbb{R}^3)^{3 \times 3}, \qquad (5.26)$$
$$\forall q \in (2, \infty): \qquad \Gamma_0^\lambda \in \mathrm{L}^q(\mathrm{B}^R)^{3 \times 3}, \qquad (5.27)$$
$$\forall q \in [1, 3/2): \qquad \partial_j \Gamma_0^\lambda \in \mathrm{L}_{\mathrm{loc}}^q(\mathbb{R}^3)^{3 \times 3}, \qquad (5.28)$$
$$\forall q \in (4/3, \infty): \qquad \partial_j \Gamma_0^\lambda \in \mathrm{L}^q(\mathrm{B}^R)^{3 \times 3} \qquad (5.29)$$

for any $R > 0$ and $j = 1, 2, 3$.

Proof. Properties (5.26) and (5.28) follow from the estimates $|\Gamma_0^\lambda(x)| \le c_0 |x|^{-1}$ and $|\nabla \Gamma_0^\lambda(x)| \le c_1 |x|^{-2}$ for $0 < |x| \le \varepsilon$ (see [26, Lemma 3.2]), and from (5.25). For (5.27) and (5.29) we refer to [42, (VII.3.28) and (VII.3.33)]. □

Similarly to the steady-state Oseen fundamental solution Γ_0^λ considered above, the fundamental solution $\Gamma_{\mathrm{H}}^{\eta,\lambda}$ to the Helmholtz equation with drift term, defined in (5.12), can be simplified in the case of dimension $n = 3$. Because the Hankel function $H_\nu^{(1)}$ for $\nu = \frac{1}{2}$ is given by

$$H_{1/2}^{(1)}(z) = -i\sqrt{\frac{2}{\pi z}}\, e^{iz},$$

we have

$$\Gamma_{\mathrm{H}}^{\eta,\lambda}:\mathbb{R}^3 \smallsetminus \{0\} \to \mathbb{C}, \qquad \Gamma_{\mathrm{H}}^{\eta,\lambda}(x) = \frac{1}{4\pi|x|}\, e^{i\sqrt{-\mu}|x|-\frac{\lambda}{2}x_1} \tag{5.30}$$

with $\mu = (\lambda/2)^2 + i\eta$ as above. The following global pointwise estimates are now a direct consequence of Lemma 5.1.1.

Lemma 5.1.9. *Let* $n = 3$, $\eta_0 > 0$ *and* $\lambda \in \mathbb{R}$. *Then*

$$|\Gamma_{\mathrm{H}}^{\eta,\lambda}(x)| \le C_{52}|x|^{-1}\, e^{-C_{45}|\eta|^{\frac{1}{2}}|x|}, \tag{5.31}$$

$$|\nabla\Gamma_{\mathrm{H}}^{\eta,\lambda}(x)| \le C_{52}\big(|x|^{-2} + |\eta|^{\frac{1}{2}}|x|^{-1}\big)\, e^{-C_{45}|\eta|^{\frac{1}{2}}|x|}, \tag{5.32}$$

for all $\eta \in \mathbb{R}$ *with* $|\eta| > \eta_0$ *and* $x \in \mathbb{R}^3 \smallsetminus \{0\}$. *Here,* C_{45} *is the constant from Lemma 5.1.1, and* $C_{52} = C_{52}(\lambda,\eta_0) > 0$.

Proof. By Lemma 5.1.1 we have

$$\left|e^{i\sqrt{-\mu}|x|-\frac{\lambda}{2}x_1}\right| \le e^{-\operatorname{Im}(\sqrt{-\mu})|x|+\frac{|\lambda|}{2}|x|} \le e^{-C_{45}|\eta|^{\frac{1}{2}}|x|}.$$

Therefore, we directly conclude (5.31) from (5.30). Computing derivatives of (5.30) and employing this estimate again, we further deduce

$$|\nabla\Gamma_{\mathrm{H}}^{\eta,\lambda}(x)| \le c_0\big(|x|^{-2} + |x|^{-1}(|\sqrt{-\mu}| + |\lambda|)\big)\, e^{-C_{45}|\eta|^{\frac{1}{2}}|x|},$$

which implies (5.32) by using $|\lambda| \le 2|\sqrt{-\mu}| \le c_1|\eta|^{\frac{1}{2}}$ for $|\eta| \ge \eta_0$. $\qquad\square$

5.2 Time-Periodic Fundamental Solutions

As mentioned before, the idea to use the concept of time-periodic fundamental solutions in the framework of the Navier–Stokes equations is quite new and goes back to [76, 21]. Since a time-periodic problem in the n-dimensional whole space $\mathbb{R}^n$ can be seen as a problem on the locally compact abelian group $G = \mathbb{T} \times \mathbb{R}^n$, where a convolution is available, it seems natural to search for a solution formula analogue to (5.3) or (5.5) and thus for a fundamental solution in the time-periodic setting. The introduction of such a time-periodic fundamental solution to the linearized Navier–Stokes equations

$$\begin{cases} \partial_t u - \Delta u - \lambda \partial_1 u + \nabla \mathfrak{p} = f & \text{in } \mathbb{T} \times \mathbb{R}^n, \\ \operatorname{div} u = 0 & \text{in } \mathbb{T} \times \mathbb{R}^n \end{cases} \tag{5.33}$$

is the first goal of this chapter. Again we let $\lambda \in \mathbb{R}$, that is, we consider both the Stokes case ($\lambda = 0$) and the Oseen case ($\lambda \neq 0$) simultaneously. For the rest of this chapter, the time period $\mathcal{T} > 0$ defining the considered torus group $\mathbb{T} = \mathbb{R}/\mathcal{T}\mathbb{Z}$ remains fixed.

As in the steady-state case, the fundamental solution will be a tensor-valued distribution that decomposes into a velocity fundamental solution and a pressure fundamental solution. While the pressure fundamental solution will more or less be given as in the steady-state case (defined in (5.10)), the velocity fundamental solution will decompose into two parts: A *steady-state* part that coincides with the velocity fundamental solution of the steady-state problem (5.4), and a second *purely periodic* part, which will be defined by means of the Fourier transform on the group $G = \mathbb{T} \times \mathbb{R}^n$. After having found suitable representation formulas for the time-periodic fundamental solutions, we further analyze their purely periodic parts and show how they provide *a priori* estimates. At the end of the section, we have a closer look at the three-dimensional case.

5.2.1 The Time-Periodic Stokes and Oseen Fundamental Solutions

In the following, we introduce a time-periodic fundamental solution to the linearized Navier–Stokes equations (5.33). By analogy to (5.6) and (5.7), a fundamental solution Φ to (5.33) is a tensor field

$$\Phi := \begin{pmatrix} \Gamma_{11}^\lambda & \cdots & \Gamma_{1n}^\lambda \\ \vdots & \ddots & \vdots \\ \Gamma_{n1}^\lambda & \cdots & \Gamma_{nn}^\lambda \\ \gamma_1 & \cdots & \gamma_n \end{pmatrix} \in \mathscr{S}'(G)^{(n+1)\times n} \tag{5.34}$$

that satisfies

$$\begin{cases} \partial_t \Gamma_{j\ell}^\lambda - \Delta \Gamma_{j\ell}^\lambda - \lambda \partial_1 \Gamma_{j\ell}^\lambda + \partial_j \gamma_\ell = \delta_{j\ell} \delta_G, \\ \partial_h \Gamma_{h\ell}^\lambda = 0 \end{cases} \tag{5.35}$$

in the sense of $\mathscr{S}'(G)$ distributions for $j, \ell = 1, \ldots, n$. A solution to the time-periodic system (5.33) is then given by

$$\begin{pmatrix} u \\ \mathfrak{p} \end{pmatrix} := \Phi * f \tag{5.36}$$

provided $f \in \mathscr{S}(G)$, where the componentwise convolution is taken over the group G. As in the stationary case considered in Section 5.1, the fundamental solution Φ consists of two parts: The *velocity fundamental solution* $\Gamma^\lambda = (\Gamma^\lambda_{j\ell})^n_{j,\ell=1} \mathscr{S}'(G)^{n \times n}$ and the *pressure fundamental solution* $\gamma = (\gamma_\ell)^n_{\ell=1} \in \mathscr{S}'(\hat{G})^n$ such that

$$u = \Gamma^\lambda * f, \qquad \mathfrak{p} = \gamma * f.$$

In the following, we shall identify a fundamental solution Φ to (5.33) as the sum of a fundamental solution to the corresponding steady-state system (5.4) and a second purely periodic fundamental solution, which we introduce by means of the Fourier transform $\mathscr{F}_G$ on the group $G = \mathbb{T} \times \mathbb{R}^n$. More specifically, this fundamental solution is given as the inverse Fourier transform of a function on $\hat{G} = \mathbb{Z} \times \mathbb{R}^n$. Recall the definition of Γ^S, Γ^O and γ_0 from (5.8), (5.9) and (5.10), respectively.

Theorem 5.2.1. *Let $n \geq 2$ and $\lambda \in \mathbb{R}$. Put*

$$\Gamma^\lambda_0 := \begin{cases} \Gamma^S & \text{if } \lambda = 0 \quad \text{(Stokes case)}, \\ \Gamma^O & \text{if } \lambda \neq 0 \quad \text{(Oseen case)}. \end{cases}$$

Then the elements of $\mathscr{S}'(G)$ given by

$$\Gamma^\lambda := \Gamma^\lambda_0 \otimes 1_{\mathbb{T}} + \Gamma^\lambda_\perp, \tag{5.37}$$

$$\gamma := \gamma_0 \otimes \delta_{\mathbb{T}}, \tag{5.38}$$

with

$$\Gamma^\lambda_\perp := \mathscr{F}^{-1}_G \left[\frac{1 - \delta_{\mathbb{Z}}(k)}{|\xi|^2 + i\left(\frac{2\pi}{\mathcal{T}}k - \lambda\xi_1\right)} \left(I - \frac{\xi \otimes \xi}{|\xi|^2} \right) \right] \in \mathscr{S}'(G)^{n \times n} \tag{5.39}$$

define a fundamental solution $\Phi \in \mathscr{S}'(G)^{(n+1) \times n}$ to (5.35) of the form (5.34).

Proof. At first, note that the function

$$M: \hat{G} \to \mathbb{C}, \quad M(k, \xi) := \frac{1 - \delta_{\mathbb{Z}}(k)}{|\xi|^2 + i\left(\frac{2\pi}{\mathcal{T}}k - \lambda\xi_1\right)} \tag{5.40}$$

is an element of $L^\infty(\widehat{G})$. Therefore, $\Gamma_\perp^\lambda$ is a well-defined tempered distribution in $\mathscr{S}'(G)^{n\times n}$. Due to the identity $\mathscr{F}_{\mathbb{R}^n}[\gamma_0] = -i\frac{\xi}{|\xi|^2}$, we obtain

$$\mathscr{F}_G[\nabla\gamma] = \mathscr{F}_{\mathbb{R}^n}[\nabla\gamma_0] \otimes \mathscr{F}_{\mathbb{T}}[\delta_{\mathbb{T}}] = \frac{\xi\otimes\xi}{|\xi|^2}\cdot 1_\mathbb{Z}.$$

By (5.7) we also have

$$\left(|\xi|^2 - i\lambda\xi_1\right)\mathscr{F}_{\mathbb{R}^n}[\Gamma_0^\lambda] = \mathscr{F}_{\mathbb{R}^n}\left[[-\Delta - \lambda\partial_1]\Gamma_0^\lambda\right] = \mathscr{F}_{\mathbb{R}^n}\left[\delta_{\mathbb{R}^n} - \nabla\gamma_0\right] = \left(I - \frac{\xi\otimes\xi}{|\xi|^2}\right).$$

Since $k\,\mathscr{F}[1_\mathbb{T}](k) = k\delta_\mathbb{Z}(k) = 0$, we deduce

$$\left(|\xi|^2 + i\frac{2\pi}{\mathcal{T}}k - i\lambda\xi_1\right)\mathscr{F}_G[\Gamma_0^\lambda \otimes 1_\mathbb{T}] = \left(I - \frac{\xi\otimes\xi}{|\xi|^2}\right)\delta_\mathbb{Z}(k).$$

This finally leads us to

$$\left(|\xi|^2 + i\frac{2\pi}{\mathcal{T}}k - i\lambda\xi_1\right)\mathscr{F}_G[\Gamma^\lambda] + \mathscr{F}_G[\nabla\gamma] = I.$$

An application of inverse Fourier transform to this equality and the fact that $\operatorname{div}\Gamma_0^\lambda = \operatorname{div}\Gamma_\perp^\lambda = 0$ let us conclude that (Γ^λ, γ) is a fundamental solution to (5.35). $\qquad\square$

5.2.2 Pointwise Estimates and Integrability

Since properties of the steady-state fundamental solution $\Phi_0 = (\Gamma_0^\lambda, \gamma_0)$ are well known (see Section 5.1), the analysis of the fundamental solution $\Phi = (\Gamma^\lambda, \gamma)$ from Theorem 5.2.1 reduces to the investigation of the purely periodic part $\Gamma_\perp^\lambda$ of the velocity fundamental solution given in (5.39). Therefore, we establish decay estimates and integrability properties of $\Gamma_\perp^\lambda$ in the following. These results are new and were published in [21].

Let us begin with the spatial decay estimate of the fundamental solution. The idea is to express $\Gamma_\perp^\lambda$ as a Fourier series on $\mathbb{T}$ and to identify the Fourier coefficients as convolutions in $\mathbb{R}^n$ of the type studied in Lemma 5.1.5.

Theorem 5.2.2. *Let $n \geq 2$, $\lambda \in \mathbb{R}$ and $r \in [1, \infty)$. For all $m \in \mathbb{N}_0$ and $\varepsilon > 0$, there exists a constant $C_{53}(n, \lambda, r, m, \varepsilon) > 0$ such that*

$$\forall |x| \geq \varepsilon: \quad \|\nabla^m \Gamma_\perp^\lambda(\cdot, x)\|_{L^r(\mathbb{T})} \leq C_{53}|x|^{-n-m}. \tag{5.41}$$

Proof. Recall Γ_{L} defined in (5.2) and $\Gamma_{\mathrm{H}}^{\frac{2\pi}{T}k,\lambda}$ defined in (5.12) (with $\eta =$ $\frac{2\pi}{T}k$). Since $\partial_j\partial_\ell[\Gamma_{\mathrm{L}} * \Gamma_{\mathrm{H}}^{\frac{2\pi}{T}k,\lambda}]$ is locally integrable by Lemma 5.1.5 and satisfies the decay estimate (5.22), it is a tempered distribution on $\mathbb{R}^n$. Therefore, we may apply the Fourier transform to this distribution. Then identity (5.19) yields

$$\mathscr{F}_{\mathbb{R}^n}\big[\partial_j\partial_\ell[\Gamma_{\mathrm{L}} * \Gamma_{\mathrm{H}}^{\frac{2\pi}{T}k,\lambda}]\big](\xi) = \frac{\xi_j\xi_\ell}{|\xi|^2}\frac{1}{|\xi|^2 + i\big(\frac{2\pi}{T}k - \lambda\xi_1\big)}$$

and, in particular,

$$\mathscr{F}_{\mathbb{R}^n}\big[\Delta[\Gamma_{\mathrm{L}} * \Gamma_{\mathrm{H}}^{\frac{2\pi}{T}k,\lambda}]\big](\xi) = \frac{1}{|\xi|^2 + i\big(\frac{2\pi}{T}k - \lambda\xi_1\big)}.$$

Hence the definition of $\Gamma_\perp^\lambda$ in (5.39) yields

$$\mathrm{D}^\alpha\Gamma_{\perp,j\ell}^\lambda = \mathscr{F}_{\mathbb{T}}^{-1}\big[\big(1 - \delta_{\mathbb{Z}}(k)\big)\big[\delta_{j\ell}\Delta - \partial_j\partial_\ell\big]\mathrm{D}^\alpha[\Gamma_{\mathrm{L}} * \Gamma_{\mathrm{H}}^{\frac{2\pi}{T}k,\lambda}]\big]. \tag{5.42}$$

This representation allows us to derive (5.41) from Lemma 5.1.5. Clearly, since $\mathbb{T}$ is a finite measure space, it suffices to consider $r \in [2, \infty)$. Then the Hölder conjugate $r' = r/(r-1)$ satisfies $r' \leq 2$. Hence the Hausdorff–Young inequality in combination with estimate (5.22) yields

$$\|\mathrm{D}^\alpha\Gamma_{\perp,j\ell}^\lambda(\cdot,x)\|_{\mathrm{L}^r(\mathbb{T})}$$

$$\leq \left(\sum_{k\in\mathbb{Z}}\Big|\big(1 - \delta_{\mathbb{Z}}(k)\big)\big[\delta_{j\ell}\Delta - \partial_j\partial_\ell\big]\mathrm{D}^\alpha[\Gamma_{\mathrm{L}} * \Gamma_{\mathrm{H}}^{\frac{2\pi}{T}k,\lambda}](x)\Big|^{r'}\right)^{\frac{1}{r'}}$$

$$\leq c_0|x|^{-n-|\alpha|}\left(\sum_{k\in\mathbb{Z}\setminus\{0\}}|k|^{-r'}\right)^{\frac{1}{r'}}.$$

Since $r' > 1$, the remaining series converges and we arrive at (5.41). $\qquad\square$

Next, we derive integrability properties of the purely periodic fundamental solution $\Gamma_\perp^\lambda$. For this, we express it as a Fourier multiplier operator applied to a function in an appropriate L^q space. In order to show that these multipliers are in fact $\mathrm{L}^q(G)$ multipliers, we employ the Transference Principle (Theorem 2.2.2) to connect them to the multipliers considered in Lemma A.3.7 in the Euclidean setting.

Theorem 5.2.3. *Let $n \geq 2$ and $\lambda \in \mathbb{R}$. Then*

$$\forall q \in \left(1, \frac{n+2}{n}\right): \quad \Gamma_\perp^\lambda \in \mathrm{L}^q(G)^{n\times n}, \tag{5.43}$$

$$\forall q \in \left[1, \frac{n+2}{n+1}\right): \quad \partial_j\Gamma_\perp^\lambda \in \mathrm{L}^q(G)^{n\times n} \quad (j = 1,\dots,n). \tag{5.44}$$

Proof. We start with the derivation of (5.43). A reformulation of equation (5.39) leads to the representation

$$
\Gamma^\lambda_{\perp,j\ell} = \mathscr{F}_G^{-1}\left[\left(\delta_{j\ell}\frac{\xi_m\xi_m}{|\xi|^2} - \frac{\xi_j\xi_\ell}{|\xi|^2}\right)\frac{1-\delta_\mathbb{Z}(k)}{|\xi|^2 + i(\frac{2\pi}{\mathcal{T}}k - \lambda\xi_1)}\right]
$$
$$
= \left[-\delta_{j\ell}(\mathfrak{R}_m\mathfrak{R}_m) + \mathfrak{R}_j\mathfrak{R}_\ell\right]\circ\mathscr{F}_G^{-1}\left[M_0\,\mathscr{F}_G\big[\mathscr{F}_G^{-1}(\mathcal{K})\big]\right],
$$

where $\mathfrak{R}_j$ denotes the Riesz transform (compare (A.81)), and we have set

$$
M_0\colon\widehat{G}\to\mathbb{C}, \quad M_0(k,\xi) := \frac{(1-\delta_\mathbb{Z}(k))|k|^{\frac{2}{n+2}}(1+|\xi|^2)^{\frac{n}{n+2}}}{|\xi|^2 + i(\frac{2\pi}{\mathcal{T}}k - \lambda\xi_1)} \tag{5.45}
$$

and

$$
\mathcal{K}\colon\widehat{G}\to\mathbb{C}, \quad \mathcal{K}(k,\xi) := (1-\delta_\mathbb{Z}(k))|k|^{-\frac{2}{n+2}}(1+|\xi|^2)^{-\frac{n}{n+2}}. \tag{5.46}
$$

By Proposition A.3.4, we have $\mathfrak{R}_j \in \mathcal{L}(\mathrm{L}^r(G))$ for all $r \in (1,\infty)$. Moreover, we obtain $M_0 = m_0|_{\mathbb{Z}\times\mathbb{R}^n}$ with m_0 defined in (A.83), where $\theta = \frac{2}{n+2}$. Since m_0 is a continuous $\mathrm{L}^r(\mathbb{R}\times\mathbb{R}^n)$ multiplier by Lemma A.3.7, an application of the Transference Principle (Theorem 2.2.2) implies that M_0 is an $\mathrm{L}^r(G)$ multiplier for all $r \in (1,\infty)$. Hence we conclude $\Gamma^\lambda_\perp \in \mathrm{L}^q(G)$ if $\mathscr{F}_G^{-1}(\mathcal{K}) \in \mathrm{L}^q(G)$. Since we have $\mathscr{F}_G^{-1}(\mathcal{K}) = \varphi_\alpha \otimes \psi_\beta$ with $\alpha = \frac{2}{n+2}$ and $\beta = \frac{2n}{n+2}$, and φ_α and ψ_β defined in (A.78) and (A.79), respectively, Proposition A.3.1 and Proposition A.3.2 imply $\mathscr{F}_G^{-1}(\mathcal{K}) \in \mathrm{L}^r(G)$ for all $r \in \left(1,\frac{n+2}{n}\right)$, and we have verified (5.43).

In order to show (5.44), we proceed in a similar way. As above, we obtain the identity

$$
\partial_h\Gamma^\lambda_{\perp,j\ell} = \left[-\delta_{j\ell}(\mathfrak{R}_m\mathfrak{R}_m) + \mathfrak{R}_j\mathfrak{R}_\ell\right]\circ\mathscr{F}_G^{-1}\left[M_h\,\mathscr{F}_G\big[\mathscr{F}_G^{-1}(\mathcal{J})\big]\right]
$$

for $h = 1,\dots,n$, where

$$
M_h\colon\widehat{G}\to\mathbb{C}, \quad M_h(k,\xi) := \frac{(1-\delta_\mathbb{Z}(k))|k|^{\frac{1}{n+2}}(1+|\xi|^2)^{\frac{n}{2(n+2)}}i\xi_h}{|\xi|^2 + i(\frac{2\pi}{\mathcal{T}}k - \lambda\xi_1)} \tag{5.47}
$$

and

$$
\mathcal{J}\colon\widehat{G}\to\mathbb{C}, \quad \mathcal{J}(k,\xi) := (1-\delta_\mathbb{Z}(k))|k|^{-\frac{1}{n+2}}(1+|\xi|^2)^{-\frac{n}{2(n+2)}}. \tag{5.48}
$$

Then we have $M_h = m_h|_{\mathbb{Z} \times \mathbb{R}^n}$ with m_h defined in (A.84), where $\theta = \frac{1}{n+2}$. Using the Transference Principle (Theorem 2.2.2) and Lemma A.3.7 again, we conclude that M_h is an $L^r(G)$ multiplier for all $r \in (1, \infty)$. Moreover, we see $\mathscr{F}_G^{-1}[\mathcal{J}] = \varphi_\alpha \otimes \psi_\beta$ with $\alpha = \frac{1}{n+2}$ and $\beta = \frac{n}{n+2}$. Arguing as above, we conclude $\mathscr{F}_G^{-1}[\mathcal{J}] \in L^q(G)$ and thus $\partial_h \Gamma_\perp^\lambda \in L^q(G)$ for all $q \in \left(1, \frac{n+2}{n+1}\right)$. In particular, this yields $\partial_h \Gamma_\perp^\lambda \in L_{loc}^1(G)$. Together with the asymptotic behavior from (5.41) for $m = 1$, this leads to $\partial_h \Gamma_\perp^\lambda \in L^1(G)$. In total, we have thus also shown (5.44) and completed the proof. $\qquad\square$

5.2.3 A priori Estimates

In the following, we demonstrate how to employ the fundamental solution in order to derive L^q estimates. We further derive unique existence of a solution to (5.33) in the associated functional framework. For the deduction of *a priori* estimates, we employ the Transference Principle (Theorem 2.2.2), which leads to the following result.

Lemma 5.2.4. *If $f \in \mathscr{S}(G)^n$ and $q \in (1, \infty)$, then $\Gamma_\perp^\lambda * f \in W_\perp^{1,2,q}(G)^n$ and there exists a polynomial $P : \mathbb{R} \to \mathbb{R}$, which only depends on n and q, such that*

$$\|\partial_t(\Gamma_\perp^\lambda * f)\|_q + \|\nabla^2(\Gamma_\perp^\lambda * f)\|_q \leq P(\lambda^2 \mathcal{T})\|f\|_q. \tag{5.49}$$

Proof. By the definition of $\Gamma_\perp^\lambda$ in (5.39), the convolution $\Gamma_\perp^\lambda * f$ can be expressed in terms of a Fourier multiplier

$$(\Gamma_\perp^\lambda * f)_j = \mathscr{F}_G^{-1}\left[M(k, \xi)\left(\delta_{j\ell} - \frac{\xi_j \xi_\ell}{|\xi|^2}\right)\mathscr{F}_G[f_\ell]\right]$$

$$= \left[-\delta_{j\ell}(\mathfrak{R}_m \mathfrak{R}_m) + \mathfrak{R}_j \mathfrak{R}_l\right] \circ \mathscr{F}_G^{-1}\left[M\mathscr{F}_G[f_\ell]\right],$$

with M given by (5.40). The property $M(0, \xi) = 0$ leads to the identity $P(\Gamma_\perp^\lambda * f) = \mathscr{F}_G^{-1}[\delta_\mathbb{Z} \mathscr{F}_G[\Gamma_\perp^\lambda * f]] = 0$. Next let us show that $\Gamma_\perp^\lambda * f \in W^{1,2,q}(\mathbb{T} \times \mathbb{R}^n)$. We have

$$\partial_r \partial_s (\Gamma_\perp^\lambda * f)_j = \left[-\delta_{jl}(\mathfrak{R}_h \mathfrak{R}_h) + \mathfrak{R}_j \mathfrak{R}_l\right] \circ \mathfrak{R}_r \mathfrak{R}_s \circ \mathscr{F}_G^{-1}\left[M_x \mathscr{F}_G[f_\ell]\right],$$

$$\partial_t(\Gamma_\perp^\lambda * f)_j = \left[-\delta_{j\ell}(\mathfrak{R}_m \mathfrak{R}_m) + \mathfrak{R}_j \mathfrak{R}_l\right] \circ \mathfrak{R}_\mathbb{T} \circ \mathscr{F}_G^{-1}\left[M_t \mathscr{F}_G[f_\ell]\right]$$

for $r, s = 1, \ldots, n$, where

$$M_x(k, \xi) = |\xi|^2 M(k, \xi), \qquad M_t(k, \xi) = -\frac{2\pi}{\mathcal{T}} k M(k, \xi)$$

and $\mathfrak{R}_j$ and $\mathfrak{R}_\mathbb{T}$ denote Riesz transforms defined in (A.81) and (2.2), respectively. Let $\kappa = \frac{2\pi}{T}$. Then we have $M_x = \widetilde{m}_{\kappa,\lambda}|_{\mathbb{Z}\times\mathbb{R}^n}$ for $\theta = 0$ and $M_t = -\widetilde{m}_{\kappa,\lambda}|_{\mathbb{Z}\times\mathbb{R}^n}$ for $\theta = 1$ with $\widetilde{m}_{\kappa,\lambda}$ defined in (A.87). By the Transference Principle (Theorem 2.2.2) and Lemma A.3.10, we thus conclude that M_x and M_t are $\mathrm{L}^q(G)$ multipliers for any $q \in (1, \infty)$. Together with the continuity of the Riesz transforms by Proposition A.3.4 and Proposition 2.2.4, this implies $\partial_r\partial_s(\Gamma_\perp^\lambda * f)$, $\partial_t(\Gamma_\perp^\lambda * f) \in \mathrm{L}^q(G)$ and the estimate (5.49) due to estimate (A.90). By $\mathcal{P}(\Gamma_\perp^\lambda * f) = 0$ and Poincaré's inequality, this further yields $\Gamma_\perp^\lambda * f \in \mathrm{L}^q(G)$ and consequently $\Gamma_\perp^\lambda * f \in \mathrm{W}_\perp^{1,2,q}(G)$. $\qquad\square$

In order to show that $\Gamma_\perp^\lambda * f$ is the unique solution to (5.33) for purely periodic f, we utilize the following uniqueness statement, which is proved by an application of the Fourier transform on G.

Lemma 5.2.5. *Let $(u, \mathfrak{p}) \in \mathscr{S}'(G)^{n+1}$ be a solution to (5.33) for the right-hand side $f = 0$. Then, $\mathcal{P}u$ is a polynomial in each component, $\mathcal{P}_\perp u = 0$, and $\mathfrak{p} \in \mathrm{L}_{\mathrm{loc}}^1(G)$ such that $\mathfrak{p}(t, \cdot)$ is a polynomial for almost each $t \in \mathbb{T}$.*

Proof. An application of the Fourier transform on G to $(5.33)_1$ yields

$$\left(i\frac{2\pi}{T}k + |\xi|^2 - i\lambda\xi_1\right)\widehat{u} + i\xi\widehat{\mathfrak{p}} = 0$$

with $\widehat{u} := \mathscr{F}_G[u]$ and $\widehat{\mathfrak{p}} := \mathscr{F}_G[\mathfrak{p}]$. Multiplying this equation with $i\xi$ and using $\mathrm{div}\, u = 0$, we obtain $-|\xi|^2\widehat{\mathfrak{p}} = 0$, so that $\mathrm{supp}\,\widehat{\mathfrak{p}} \subset \mathbb{Z} \times \{0\}$. Then, the above equation yields

$$\mathrm{supp}\left[\left(i\frac{2\pi}{T}k + |\xi|^2 - i\lambda\xi_1\right)\widehat{u}\right] = \mathrm{supp}\left[-i\xi\widehat{\mathfrak{p}}\right] \subset \mathbb{Z} \times \{0\}.$$

Because the only zero of $(k, \xi) \mapsto (i\frac{2\pi}{T}k + |\xi|^2 - i\lambda\xi_1)$ is $(k, \xi) = (0, 0)$, we conclude $\mathrm{supp}\,\widehat{u} \subset \{(0, 0)\}$. Thus we obtain $\mathcal{P}_\perp u = 0$ and that $\mathcal{P}u$ is a polynomial in each component. Now $(5.33)_1$ leads to $\mathfrak{p} \in \mathrm{L}_{\mathrm{loc}}^1(G)$. As above, this implies $\mathrm{supp}\,\mathscr{F}_{\mathbb{R}^n}[\mathfrak{p}(t, \cdot)] \subset \{0\}$, so that $\mathfrak{p}(t, \cdot)$ is a polynomial for almost all $t \in \mathbb{T}$. $\qquad\square$

Now we show existence of a unique solution to (5.33) when the right-hand side $f \in \mathrm{L}^q(G)^n$ is purely periodic. We first obtain a solution for $f \in \mathscr{S}(G)^n$ by means of the fundamental solution. Due to the *a priori* estimate from Lemma 5.2.4, we can then employ a standard density argument to extend this to general $f \in \mathrm{L}^q(G)^n$ with $\mathcal{P}f = 0$.

Theorem 5.2.6. *Let $q \in (1, \infty)$, $\lambda \in \mathbb{R}$ and $f \in L^q_\perp(G)$. Then there exists a solution $(u, \mathfrak{p})$ to (5.33) such that $u \in W^{1,2,q}_\perp(G)^3$, $\nabla \mathfrak{p} \in L^q_\perp(G)$, which satisfies*

$$\|\partial_t u\|_q + \|\nabla^2 u\|_q + \|\lambda \partial_1 u\|_q + \|\nabla \mathfrak{p}\|_q \leq P(\lambda^2 \mathcal{T}) \|f\|_q, \tag{5.50}$$

$$\|u\|_{1,2,q} + \|\nabla \mathfrak{p}\|_q \leq C_{54} \|f\|_q \tag{5.51}$$

for a polynomial $P : \mathbb{R} \to \mathbb{R}$, which only depends on n and q, and some constant $C_{54} = C_{54}(n, q, \lambda, \mathcal{T}) > 0$. If $(u_1, \mathfrak{p}_1) \in \mathscr{S}'(G)^3 \times \mathscr{S}'(G)$ is another solution to (5.33) with $\mathcal{P} u_1 = 0$, then $u = u_1$ and $\nabla \mathfrak{p} = \nabla \mathfrak{p}_1$.

Proof. This theorem was originally proved in [74]. We give a slightly different proof here. At first, consider $f \in \mathscr{S}(G)^n$ with $\mathcal{P} f = 0$, and define $u := \Gamma^\lambda * f$ as well as $\mathfrak{p} := \gamma * f$, where Γ^λ and γ were defined in (5.37) and (5.38). Then $(u, \mathfrak{p})$ is a solution to (5.33) due to Theorem 5.2.1. Since $\mathcal{P} f = 0$, we conclude $(\Gamma^\lambda_0 \otimes 1) * f = 0$, so that $u = \Gamma^\lambda_\perp * f$ and $u \in W^{1,2,q}_\perp(G)^n$ by Lemma 5.2.4. The identity

$$\partial_j \mathfrak{p} = \mathscr{F}_G^{-1}\left[\left(\mathscr{F}_{\mathbb{R}^n}[\partial_j \gamma_{0,\ell}] \otimes \mathscr{F}_\mathbb{T}[\delta_\mathbb{Z}]\right) \mathscr{F}_G[f_\ell]\right] = \mathscr{F}_G^{-1}\left[\frac{\xi_j \xi_\ell}{|\xi|^2} \mathscr{F}_G[f_\ell]\right] = -\mathfrak{R}_j \mathfrak{R}_\ell f_\ell$$

implies $\nabla \mathfrak{p} \in L^q(G)^n$ by Proposition A.3.4 and $\|\nabla \mathfrak{p}\|_q \leq c_0 \|f\|_q$, where c_0 is independent of λ and $\mathcal{T}$. Due to this estimate, (5.49) and the identity $\lambda \partial_1 u = \partial_t u - \Delta u + \nabla \mathfrak{p} - f$, we now conclude (5.50). Since $\mathcal{P} u = 0$, we further conclude (5.51) by Poincaré's inequality. Moreover, $\mathcal{P} u = 0$ implies $\mathcal{P} \nabla \mathfrak{p} = 0$, so that $(u, \mathfrak{p})$ is a solution with the desired properties.

Now let $f \in L^q_\perp(G)^n$. Then there exists a sequence $(f_j) \subset \mathscr{S}(G)^n$ with $\mathcal{P} f_j = 0$ that converges to f in $L^q(G)^n$. By the above argument, for each $j \in \mathbb{N}$ there exists a solution $(u_j, \mathfrak{p}_j)$ with $u_j \in W^{1,2,q}_\perp(G)^n$ and $\nabla \mathfrak{p}_j \in L^q_\perp(G)^n$ to (5.33) with right-hand side f_j, and satisfying (5.50) and (5.51). Since (f_j) is a Cauchy sequence in $L^q(G)^n$, estimate (5.51) implies that (u_j) and $(\nabla \mathfrak{p}_j)$ are Cauchy sequences in $W^{1,2,q}_\perp(G)$ and $L^q_\perp(G)$ and therefore possess limits u and $\nabla \mathfrak{p}$ in the respective spaces. Then $(u, \mathfrak{p})$ is a solution to (5.33) with right-hand side f that satisfies (5.50) and (5.51).

For the uniqueness statement, it suffices to consider a solution $(u, \mathfrak{p})$ to (5.33) for right-hand side $f = 0$ such that $\mathcal{P} u = 0$. Then Lemma 5.2.5 implies $u = \mathcal{P}_\perp u = 0$, and from (5.33)$_1$ we conclude $\nabla \mathfrak{p} = 0$. This shows the uniqueness statement. $\qquad\square$

Remark 5.2.7. Note that Theorem 5.2.6 coincides with Theorem 3.2.1 in the case of the whole space $\Omega = \mathbb{R}^n$. Moreover, Theorem 5.2.6 can be used as the starting point of a proof of Theorem 3.2.1, where $\Omega \subset \mathbb{R}^n$ is an exterior domain, by means of a localization procedure. This approach was carried out in [50] to prove Theorem 3.2.1 in the case $n = 3$.

As we have seen in Subsection 3.2.2, Theorem 5.2.6 can be combined with a well-posedness result for the steady-state Stokes or Oseen problem (5.4) in order to omit the condition $\mathcal{P}f = 0$. In this way, one can show existence of a time-periodic solution for general time-periodic right-hand sides $f \in L^q(\mathbb{T} \times \mathbb{R}^n)$ together with an *a priori* estimate.

5.2.4 The Three-Dimensional Case

Again, in view of our later application to the time-periodic Navier–Stokes equations in three dimensions, let us single out the case $n = 3$ in the following. In order to derive pointwise decay estimates of the velocity field, we shall represent it via a convolution with integral kernel Γ^λ, the time-periodic velocity fundamental solution. Due to the representation (5.37) of Γ^λ, this convolution is the sum of a convolution with the steady-state part Γ_0^λ and a convolution with the purely periodic part $\Gamma_\perp^\lambda$. While convolutions of the former type were subject of Theorem 5.1.7, here we investigate convolutions with $\Gamma_\perp^\lambda$.

Theorem 5.2.8. *Let $n = 3$ and $A \in (0, \infty)$ with $A \neq 3$. Let $g \in L^\infty(\mathbb{T} \times \mathbb{R}^3)$ such that $g(t, x) \leq M(1 + |x|)^{-A}$, and let $\varepsilon > 0$. Then for any $\delta > 0$ there exists a constant $C_{55} = C_{55}(A, \lambda, \mathcal{T}, \varepsilon, \delta) > 0$ such that*

$$\forall |x| \geq \varepsilon: \quad \left| |\Gamma_\perp^\lambda| *_G g(t, x) \right| \leq C_{55} M \begin{cases} (1 + |x|)^{-3} & \text{if } A > 3, \\ (1 + |x|)^{-A+\delta} & \text{if } A < 3, \end{cases} \quad (5.52)$$

$$\forall |x| \geq \varepsilon: \quad \left| |\nabla \Gamma_\perp^\lambda| *_G g(t, x) \right| \leq C_{55} M (1 + |x|)^{-\min\{A,4\}}. \quad (5.53)$$

Proof. Let $x \in \mathbb{R}^3$, $|x| \geq \varepsilon$ and set $R := |x|/2$. Using the estimate for g, we have

$$\left| |\Gamma_\perp^\lambda| *_G g(t, x) \right| \leq M(I_1 + I_2 + I_3)$$

where

$$I_1 = \int_{B_R} \int_{\mathbb{T}} |\Gamma_\perp^\lambda(t - s, x - y)| (1 + |y|)^{-A} \, dy ds,$$

$$I_2 = \int_{B_{4R} \setminus B_R} \int_{\mathbb{T}} |\Gamma_\perp^\lambda(t - s, x - y)| (1 + |y|)^{-A} \, dy ds,$$

$$I_3 = \int_{B^{4R}} \int_{\mathbb{T}} |\Gamma_\perp^\lambda(t - s, x - y)| (1 + |y|)^{-A} \, dy ds.$$

We estimate these terms separately. Since $|y| \le R$ implies $|x - y| \ge |x| - |y| \ge |x|/2 = R \ge \varepsilon/2$, we can use (5.41) to estimate

$$I_1 \le c_0 \int_{B_R} |x - y|^{-3}(1 + |y|)^{-A}\,dy \le c_1|x|^{-3} \int_{\mathbb{R}^3} (1 + |y|)^{-A}\,dy \le c_2|x|^{-3}$$

if $A > 3$, and

$$I_1 \le c_3 \int_{B_R} |x - y|^{-3}(1 + |y|)^{-A}\,dy \le c_4|x|^{-3} \int_{B_R} |y|^{-A}\,dy \le c_5|x|^{-A}$$

if $A < 3$. To estimate I_2, we employ Hölder's inequality with $q \in (1, \frac{5}{3})$ and $q' = q/(q - 1)$, which yields

$$I_2 \le |x|^{-A}\left(\int_{\mathbb{T}} \int_{B_{4R} \setminus B_R} 1\,dyds\right)^{1/q'} \left(\int_{\mathbb{T}} \int_{B_{4R} \setminus B_R} |\Gamma_\perp^\lambda(t - s, x - y)|^q\,dyds\right)^{1/q}$$

$$\le c_6|x|^{-A}|x|^{3 - \frac{3}{q}}\|\Gamma_\perp^\lambda\|_{L^q(\mathbb{T} \times \mathbb{R}^3)}.$$

If $A > 3$, we choose $q \in (1, \frac{5}{3})$ so small that $-A + 3 - \frac{3}{q} < -3$. If $A < 3$, we choose $q \in (1, \frac{5}{3})$ so small that $-A + 3 - \frac{3}{q} < -A + \delta$. In virtue of $\Gamma_\perp^\lambda \in L^q(\mathbb{T} \times \mathbb{R}^n)$ (by (5.43)), this implies

$$I_2 \le c_7 \begin{cases} |x|^{-3} & \text{if } A > 3, \\ |x|^{-A+\delta} & \text{if } A < 3. \end{cases}$$

For I_3 we note that $|y| \ge 4R$ implies $|x - y| \ge |y| - |x| \ge |y| - |y|/2 = |y|/2 \ge 2R \ge \varepsilon$. Therefore, (5.41) yields

$$I_3 \le c_8 \int_{B^{4R}} |x - y|^{-3}(1 + |y|)^{-A}\,dy \le c_9 \int_{B^{4R}} |y|^{-3}|y|^{-A}\,dy \le c_{10}|x|^{-A}.$$

Collecting the estimates of I_1, I_2 and I_3, we obtain (5.52).

A proof of (5.53) can be given in a similar but simpler way. For the sake of completeness, we sketch it here. We have

$$\left||\nabla \Gamma_\perp^\lambda| *_G g(t, x)\right| \le M(J_1 + J_2)$$

where

$$J_1 = \int_{B_R} \int_{\mathbb{T}} |\nabla \Gamma_\perp^\lambda(t - s, x - y)|\,(1 + |y|)^{-A}\,dsdy,$$

$$J_2 = \int_{B^R} \int_{\mathbb{T}} |\nabla \Gamma_\perp^\lambda(t - s, x - y)|\,(1 + |y|)^{-A}\,dsdy.$$

Employing (5.41), we estimate

$$J_1 \leq c_{11} \int_{B_R} |x - y|^{-4}(1 + |y|)^{-A}\, dy \leq c_{12}|x|^{-4} \int_{B_R} (1 + |y|)^{-A}\, dy.$$

With the same argument as above, this yields

$$J_1 \leq c_{13} \begin{cases} |x|^{-4} & \text{if } A > 3, \\ |x|^{-A-1} & \text{if } A < 3. \end{cases}$$

Since $\nabla \Gamma_\perp^\lambda \in \mathrm{L}^1(\mathbb{T} \times \mathbb{R}^3)$ by (5.44), we further obtain

$$J_2 \leq c_{14}|x|^{-A}\|\nabla \Gamma_\perp^\lambda\|_{\mathrm{L}^1(\mathbb{T} \times \mathbb{R}^3)} \leq c_{15}|x|^{-A}$$

Collecting the estimates of J_1 and J_2, we conclude (5.53). $\qquad\square$

Note that in the case $A = 3$ it is possible to derive an estimate that contains logarithmic terms. Since we shall not need this case, we excluded it here for the sake of simplicity.

Furthermore, in the case $A < 3$ the convolution $|\Gamma_\perp^\lambda| * g$ shows a slightly worse decay rate than the optimal one $(1 + |x|)^{-A}$. Comparing the proofs of (5.52) and (5.53), one sees that this occurs since we do not have $\Gamma_\perp^\lambda \in \mathrm{L}^1(\mathbb{T} \times \mathbb{R}^n)$. However, even if one could improve this estimate and omit δ, this would not affect the results for the Navier–Stokes equations derived in Chapter 6.

5.3 The Vorticity Fundamental Solution

In Chapter 6, one of our goals is to analyze the asymptotic behavior of the vorticity associated to the time-periodic flow around a body. As mentioned beforehand, such estimates can be derived if an integral representation for the vorticity is available. Therefore, our first task in this section shall be to derive such a representation in terms of a fundamental solution. Again, the corresponding integral kernel will consist of two parts: A first part that is the vorticity fundamental solution to the corresponding steady-state problem, and a second, purely periodic part. Afterwards, we derive decay estimates and integrability properties of this fundamental solution. Note that, in the whole section, we will only consider the case $n = 3$, though, similar results can also be derived in the case of dimension $n = 2$.

5.3.1 The Vorticity Fundamental Solution

Let $(u, \mathfrak{p})$ be a solution to the time-periodic problem (5.33) for a given right-hand side $f \colon \mathbb{T} \times \mathbb{R}^3 \to \mathbb{R}^3$. In the following we focus on the velocity field u, and we want to obtain an integral formula for the associated vorticity $\operatorname{curl} u$ similar to the representation $u = \Gamma^\lambda * f$ derived previously. Using this formula, for $m = 1, 2, 3$ we formally obtain, using Einstein summation convention,

$$(\operatorname{curl} u)_m = \varepsilon_{mhj} \partial_h (\Gamma_{j\ell}^\lambda * f_\ell) = \varepsilon_{mhj} (\partial_h \Gamma_{j\ell}^\lambda) * f_\ell.$$

In virtue of (5.24), (5.42) and (5.37), we can express Γ^λ as

$$\Gamma_{j\ell}^\lambda = \left[\delta_{j\ell} \Delta - \partial_j \partial_\ell \right] \Psi^\lambda$$

with $\Psi^\lambda = \Psi_0^\lambda \otimes 1_\mathbb{T} + \Psi_\perp^\lambda$, where

$$\Psi_0^\lambda(x) = \frac{1}{4\pi\lambda} \int_0^{s(\lambda x)/2} \frac{1 - e^{-\tau}}{\tau} \, d\tau,$$

$$\Psi_\perp^\lambda(t, x) = \mathscr{F}_\mathbb{T}^{-1} \left[k \mapsto \left(1 - \delta_\mathbb{Z}(k) \right) \left[\Gamma_\mathrm{L} * \Gamma_\mathrm{H}^{\frac{2\pi}{\mathcal{T}} k, \lambda} \right](x) \right](t).$$

We thus conclude

$$(\operatorname{curl} u)_m = \varepsilon_{mhj} \left(\left[\delta_{j\ell} \partial_h \Delta - \partial_h \partial_j \partial_\ell \right] \Psi^\lambda \right) * f_\ell = \varepsilon_{mh\ell} \partial_h \Delta \Psi^\lambda * f_\ell.$$

Setting $\phi^\lambda := \Delta \Psi^\lambda$, we thus obtain the equation

$$\operatorname{curl} u(t, x) = \int_G \nabla \phi^\lambda (t - s, x - y) \wedge f(s, y) \, d(s, y). \tag{5.54}$$

This is the desired integral representation formula for the vorticity $\operatorname{curl} u$, and we call ϕ^λ the *vorticity fundamental solution*. Note that this is a slight abuse of notation because (5.54) is not a standard convolution with the kernel ϕ^λ. Nevertheless, we continue to use this name in the following. Moreover, using the identity

$$\left| \nabla \left[s(\lambda x) \right] \right|^2 = \left| \frac{|\lambda| x}{|x|} + \lambda \, \mathrm{e}_1 \right|^2 = 2|\lambda|^2 + \frac{2|\lambda|\lambda x_1}{|x|} = \frac{2|\lambda| s(\lambda x)}{|x|}, \tag{5.55}$$

one shows by a direct computation that

$$\phi^\lambda = \phi_0^\lambda \otimes 1_\mathbb{T} + \phi_\perp^\lambda \tag{5.56}$$

with

$$\phi_0^\lambda(x) := \frac{\operatorname{sgn}(\lambda)}{4\pi|x|}\,e^{-s(\lambda x)/2}, \tag{5.57}$$

$$\phi_\perp^\lambda(t,x) := \mathscr{F}_{\mathbb{T}}^{-1}\big[k \mapsto \big(1 - \delta_{\mathbb{Z}}(k)\big)\Gamma_{\mathrm{H}}^{\frac{2\pi}{T}k,\lambda}(x)\big](t). \tag{5.58}$$

By (5.19) we also have the identity

$$\phi_\perp^\lambda = \mathscr{F}_G^{-1}\left[\frac{1 - \delta_{\mathbb{Z}}(k)}{|\xi|^2 - i\lambda\xi_1 + i\frac{2\pi}{T}k}\right]. \tag{5.59}$$

Though the integral in (5.54) is not a classical convolution due to the appearing vector product, estimates can be derived in the very same way. In particular, spatial decay estimates of $\operatorname{curl} u$ given by (5.54) are available if we have sufficient pointwise information on $\nabla\phi^\lambda$. These will be derived in the following.

5.3.2 Pointwise Estimates and Integrability

After having introduced the vorticity fundamental solution ϕ^λ, we now derive corresponding integrability properties and decay estimates similar to those from Subsection 5.2.2. We begin with recalling the following pointwise estimates of the steady-state vorticity fundamental solution ϕ_0^λ defined in (5.57). The proof is a straightforward calculation.

Theorem 5.3.1. *Let $\lambda \in \mathbb{R} \setminus \{0\}$. There exists $C_{56} = C_{56}(\lambda) > 0$ such that for all $x \in \mathbb{R}^3 \setminus \{0\}$ the steady-state vorticity fundamental solution ϕ_0^λ satisfies*

$$|\phi_0^\lambda(x)| \leq C_{56}|x|^{-1}\,e^{-s(\lambda x)/2}, \tag{5.60}$$

$$|\nabla\phi_0^\lambda(x)| \leq C_{56}\big(|x|^{-2} + |x|^{-3/2}s(\lambda x)^{1/2}\big)\,e^{-s(\lambda x)/2}. \tag{5.61}$$

Proof. Estimate (5.60) is indeed trivial and follows directly from (5.57). For estimate (5.61) we take derivatives in (5.57) and obtain

$$|\nabla\phi_0^\lambda(x)| \leq c_0\big(|x|^{-2} + |x|^{-1}|\nabla[s(\lambda x)]|\big)\,e^{-\frac{1}{2}s(\lambda x)}.$$

Now (5.61) follows from the identity (5.55) and the proof is complete. $\qquad\square$

Next we show analogous estimates of the purely periodic part $\phi_\perp^\lambda$ defined in (5.58). For this purpose, we first study the following Fourier multiplier.

Let $\chi \in C^\infty(\mathbb{R})$, $0 \leq \chi \leq 1$, with $\chi(\eta) = 0$ for $|\eta| \leq \frac{1}{2}$ and $\chi(\eta) = 1$ for $|\eta| \geq 1$. For $\alpha \in \mathbb{N}_0^3$ with $|\alpha| \leq 1$, $\gamma \in (0,1)$ and $x \in \mathbb{R}^3 \setminus \{0\}$ define the function

$$m_{\alpha,x}:\mathbb{R} \to \mathbb{R}, \qquad m_{\alpha,x}(\eta) := \chi(\eta)|\eta|^\gamma D^\alpha \Gamma_{\mathrm{H}}^{\frac{2\pi}{\mathcal{T}}\eta,\lambda}(x), \qquad (5.62)$$

where $\Gamma_{\mathrm{H}}^{\eta,\lambda}$ is defined in (5.30). Employing the pointwise estimates of $\Gamma_{\mathrm{H}}^{\eta,\lambda}$ established in Lemma 5.1.9, we show that $m_{\alpha,x}$ is an $L^q(\mathbb{R})$ multiplier.

Lemma 5.3.2. *Let $\lambda \in \mathbb{R}$ and $\mathcal{T} > 0$. Let $\alpha \in \mathbb{N}_0^3$ with $|\alpha| \leq 1$, $\gamma \in (0,1)$ and $x \in \mathbb{R}^3 \setminus \{0\}$. Then $m_{\alpha,x}$ is an $L^q(\mathbb{R})$ multiplier for any $q \in (1,\infty)$, and there exist constants $C_{57} = C_{57}(\lambda, \mathcal{T}, q, \alpha, \gamma) > 0$ and $C_{58} = C_{58}(\lambda, \mathcal{T}) > 0$ such that*

$$\|\mathrm{op}_\mathbb{R}[m_{\alpha,x}]\|_{\mathcal{L}(L^q(\mathbb{R}))} \leq C_{57}|x|^{-1-|\alpha|-2\gamma} e^{-C_{58}|x|}.$$

Proof. We show the statement by an application of the Marcinkiewicz Multiplier Theorem (Theorem A.3.3). For this, we need to derive suitable estimates of $m_{\alpha,x}$ and $\eta \partial_\eta m_{\alpha,x}$.

At first, let $\alpha = 0$. From (5.31) we conclude

$$|m_{0,x}(\eta)| \leq c_0 \chi(\eta)|\eta|^\gamma |x|^{-1} e^{-C_{45}|\frac{2\pi}{\mathcal{T}}\eta|^{\frac{1}{2}}|x|} \leq c_1|x|^{-1-2\gamma} e^{-C_{45}|\frac{2\pi}{\mathcal{T}}\eta|^{\frac{1}{2}}|x|/2}$$

for $|\eta| \geq \frac{1}{2}$. Moreover, differentiating $\Gamma_{\mathrm{H}}^{\frac{2\pi}{\mathcal{T}}\eta,\lambda}$ with respect to η, we obtain

$$|\partial_\eta \Gamma_{\mathrm{H}}^{\frac{2\pi}{\mathcal{T}}\eta,\lambda}(x)| \leq c_2|\partial_\eta \sqrt{-\mu}||x||\Gamma_{\mathrm{H}}^{\frac{2\pi}{\mathcal{T}}\eta,\lambda}(x)| \leq c_3|\eta|^{-\frac{1}{2}}|x||\Gamma_{\mathrm{H}}^{\frac{2\pi}{\mathcal{T}}\eta,\lambda}(x)|,$$

so that (5.31) yields

$$|\eta \partial_\eta m_{0,x}(\eta)|$$

$$\leq |\chi'(\eta)|\eta|^{\gamma+1}\Gamma_{\mathrm{H}}^{\frac{2\pi}{\mathcal{T}}\eta,\lambda}(x)| + |\chi(\eta)\gamma|\eta|^\gamma \Gamma_{\mathrm{H}}^{\frac{2\pi}{\mathcal{T}}\eta,\lambda}(x)| + |\chi(\eta)|\eta|^{\gamma+1}\partial_\eta \Gamma_{\mathrm{H}}^{\frac{2\pi}{\mathcal{T}}\eta,\lambda}(x)|$$

$$\leq c_4\big(|\eta|^\gamma|x|^{-1} + |\eta|^{\gamma+\frac{1}{2}}\big) e^{-C_{45}|\frac{2\pi}{\mathcal{T}}\eta|^{\frac{1}{2}}|x|} \leq c_5|x|^{-1-2\gamma} e^{-C_{45}|\frac{2\pi}{\mathcal{T}}\eta|^{\frac{1}{2}}|x|/2}$$

for $|\eta| \geq \frac{1}{2}$. Collecting these estimates and utilizing $m_{0,x}(\eta) = 0$ for $|\eta| \leq \frac{1}{2}$, we have

$$|m_{0,x}(\eta)| + |\eta \partial_\eta m_{0,x}(\eta)| \leq c_6|x|^{-1-2\gamma} e^{-C_{58}|x|} \qquad (5.63)$$

with $C_{58} = \sqrt{\pi/\mathcal{T}}C_{45}/2$ for all $\eta \in \mathbb{R}$.

Next consider the case $|\alpha| = 1$, that is, $\alpha = e_j$ for some $j \in \{1,2,3\}$. Then (5.32) leads to

$$|m_{\alpha,x}(\eta)| \leq c_7 \chi(\eta)|\eta|^\gamma\big(|x|^{-2} + |\eta|^{\frac{1}{2}}|x|^{-1}\big) e^{-C_{45}|\frac{2\pi}{\mathcal{T}}\eta|^{\frac{1}{2}}|x|}$$

$$\leq c_8|x|^{-2-2\gamma} e^{-C_{45}|\frac{2\pi}{\mathcal{T}}\eta|^{\frac{1}{2}}|x|/2}$$

for $|\eta| \geq \frac{1}{2}$. Moreover, a straightforward calculation yields

$$\left|\partial_\eta \partial_j \Gamma_H^{\frac{2\pi}{\mathcal{T}}\eta,\lambda}(x)\right| \leq c_9\left(|\mu|^{-\frac{1}{2}} + |x|\right)\left|\Gamma_H^{\frac{2\pi}{\mathcal{T}}\eta,\lambda}(x)\right| \leq c_{10}\left(|\eta|^{-\frac{1}{2}} + |x|\right)\left|\Gamma_H^{\frac{2\pi}{\mathcal{T}}\eta,\lambda}(x)\right|,$$

so that we can employ Lemma 5.1.9 to estimate

$$\left|\eta\partial_\eta\partial_j m_{\alpha,x}(\eta)\right| \leq \left|\chi'(\eta)|\eta|^{\gamma+1}\partial_j \Gamma_H^{\frac{2\pi}{\mathcal{T}}\eta,\lambda}(x)\right| + \left|\chi(\eta)\gamma|\eta|^\gamma\partial_j\Gamma_H^{\frac{2\pi}{\mathcal{T}}\eta,\lambda}(x)\right|$$

$$+ \left|\chi(\eta)|\eta|^{\gamma+1}\partial_\eta\partial_j\Gamma_H^{\frac{2\pi}{\mathcal{T}}\eta,\lambda}(x)\right|$$

$$\leq c_{11}\left(|\eta|^\gamma|x|^{-2} + |\eta|^{\gamma+\frac{1}{2}}|x|^{-1} + |\eta|^{\gamma+1}\right)e^{-C_{45}|\frac{2\pi}{\mathcal{T}}\eta|^{\frac{1}{2}}|x|}$$

$$\leq c_{12}|x|^{-2-2\gamma}e^{-C_{45}|\frac{2\pi}{\mathcal{T}}\eta|^{\frac{1}{2}}|x|/2}$$

for $|\eta| \geq \frac{1}{2}$. Collecting these estimates and utilizing $m_{\alpha,x}(\eta) = 0$ for $|\eta| \leq \frac{1}{2}$, we have

$$\left|m_{\alpha,x}(\eta)\right| + \left|\eta\partial_\eta m_{\alpha,x}(\eta)\right| \leq c_{13}|x|^{-2-2\gamma}e^{-C_{58}|x|} \tag{5.64}$$

with $C_{58} = \sqrt{\pi/\mathcal{T}}\,C_{45}/2$ as above.

By the Marcinkiewicz Multiplier Theorem (Theorem A.3.3), the assertion is a direct consequence of (5.63) and (5.64). $\qquad\square$

To establish pointwise estimates of $\phi_\perp^\lambda$, we express it by means of a Fourier multiplier that can be controlled with the previous lemma and the Transference Principle (Theorem 2.2.2).

Theorem 5.3.3. *For all* $\gamma \in (0,1)$, $q \in [1, \frac{1}{1-\gamma})$ *and* $x \in \mathbb{R}^3 \setminus \{0\}$ *the time-periodic vorticity fundamental solution satisfies the estimates*

$$\|\phi_\perp^\lambda(\cdot,x)\|_{L^q(\mathbb{T})} \leq C_{59}|x|^{-(1+2\gamma)}e^{-C_{58}|x|}, \tag{5.65}$$

$$\|\nabla\phi_\perp^\lambda(\cdot,x)\|_{L^q(\mathbb{T})} \leq C_{59}|x|^{-(2+2\gamma)}e^{-C_{58}|x|} \tag{5.66}$$

for constants $C_{59} = C_{59}(\lambda,\mathcal{T},q,\gamma) > 0$ *and* $C_{58} = C_{58}(\lambda,\mathcal{T}) > 0$.

Proof. First of all, note that it suffices to consider the case $q > 1$ since $\mathbb{T}$ is a finite measure space. We fix $x \in \mathbb{R}^3$, $x \neq 0$. Due to formula (5.58), we have

$$D^\alpha\phi_\perp^\lambda(\cdot,x) = \mathscr{F}_\mathbb{T}^{-1}\left[M_{\alpha,x}\mathscr{F}_\mathbb{T}[\varphi_\gamma]\right] \tag{5.67}$$

with

$$M_{\alpha,x}(k) := \left(1 - \delta_\mathbb{Z}(k)\right)|k|^\gamma D^\alpha\Gamma_H^{\frac{2\pi}{\mathcal{T}}k,\lambda}(x)$$

and $\varphi_\gamma = \mathscr{F}_{\mathbb{T}}^{-1}\big[k \mapsto \big(1 - \delta_{\mathbb{Z}}(k)\big)|k|^{-\gamma}\big]$. First, note that $M_{\alpha,x} = m_{\alpha,x}|_{\mathbb{Z}}$ for $m_{\alpha,x}$ defined in (5.62). Since $m_{\alpha,x}$ is continuous and an $L^q(\mathbb{R})$ multiplier by Lemma 5.3.2, the Tansference Principle (Theorem 2.2.2) implies that $M_{\alpha,x}$ is an $L^q(\mathbb{T})$ multiplier for any $q \in (1,\infty)$ and

$$\|\mathrm{op}_{\mathbb{T}}[M_{\alpha,x}]\|_{\mathcal{L}(L^q(\mathbb{T}))} \leq c_0 |x|^{-1-|\alpha|-2\gamma}\, e^{-C_{58}|x|}.$$

Moreover, Proposition A.3.1 yields $\varphi_\gamma \in L^q(\mathbb{T})$ provided $q < 1/(1-\gamma)$. Finally, representation formula (5.67) leads to

$$\|\mathrm{D}^\alpha \phi_\perp^\lambda(\cdot,x)\|_{L^q(\mathbb{T})} \leq \|\mathrm{op}_{\mathbb{T}}[M_{\alpha,x}]\|_{\mathcal{L}(L^q(\mathbb{T}))}\, \|\varphi_\gamma\|_{L^q(\mathbb{T})} \leq c_1 |x|^{-1-|\alpha|-2\gamma}\, e^{-C_{58}|x|},$$

which finishes the proof. □

Analogously to the proof of Theorem 5.2.3, we can establish the following integrability properties by using the representation formula (5.59).

Theorem 5.3.4. *Let $\lambda \in \mathbb{R}$. Then*

$$\forall q \in \left[1, \frac{5}{3}\right): \quad \phi_\perp^\lambda \in L^q(G), \tag{5.68}$$

$$\forall q \in \left[1, \frac{5}{4}\right): \quad \nabla\phi_\perp^\lambda \in L^q(G)^3. \tag{5.69}$$

Proof. Reformulating (5.59), we obtain

$$\phi_\perp^\lambda = \mathscr{F}_G^{-1}\Big[M_0\,\mathscr{F}_G\big[\mathscr{F}_G^{-1}(\mathcal{K})\big]\Big], \qquad \partial_h\phi_\perp^\lambda = \mathscr{F}_G^{-1}\Big[M_h\,\mathscr{F}_G\big[\mathscr{F}_G^{-1}(\mathcal{J})\big]\Big]$$

for $h = 1,2,3$, where M_0, $\mathcal{K}$, M_h and $\mathcal{J}$ are defined in (5.45), (5.46), (5.47) and (5.48), respectively. As shown in the proof of Theorem 5.2.3, M_0 and M_h are $L^r(G)$ multipliers for all $r \in (1,\infty)$. Moreover, $\mathscr{F}_G^{-1}[\mathcal{K}] \in L^q(G)$ for $q \in (1,\frac{5}{3})$ and $\mathscr{F}_G^{-1}[\mathcal{J}] \in L^q(G)$ for $q \in (1,\frac{4}{3})$. Therefore, the above representations yield (5.68) and (5.69) except for the case $q = 1$. However, these integrability properties yield $\phi_\perp^\lambda, \partial_h\phi_\perp^\lambda \in L^1_{\mathrm{loc}}(G)$, and the case $q = 1$ follows from the pointwise estimates established in Theorem 5.3.3. □

Remark 5.3.5. Observe that in Theorem 5.3.3 and Theorem 5.3.4 the case $\lambda = 0$ is not excluded. Therefore, these results also applicable in the analysis of time-periodic Stokes flow.

Now we have completed the study of fundamental solutions for our purposes. In the next chapter we apply the presented results in order to analyze the spatially asymptotic behavior of the velocity and vorticity fields associated to a time-periodic Navier–Stokes flow.

6 Spatial Decay of Time-Periodic Solutions to the Navier–Stokes Equations

In this chapter, we study the asymptotic behavior of a time-periodic Navier–Stokes flow past a body. More precisely, we derive spatial decay estimates of solutions $(u, \mathfrak{p})$ to the time-periodic Navier–Stokes equations

$$\begin{cases} \partial_t u - \Delta u - \lambda \partial_1 u + u \cdot \nabla u + \nabla \mathfrak{p} = f & \text{in } \mathbb{T} \times \Omega, \\ \operatorname{div} u = 0 & \text{in } \mathbb{T} \times \Omega, \\ \lim_{|x| \to \infty} u(t, x) = 0 & \text{for } t \in \mathbb{T} \end{cases} \tag{6.1}$$

in a three-dimensional exterior domain $\Omega \subset \mathbb{R}^3$ with Reynolds number $\lambda \neq 0$ and a given right-hand side f. Note that it is not necessary to specify the boundary values of u at $\partial\Omega$ in this framework since we study spatial asymptotic properties of solutions that exist by assumption. Clearly, to ensure existence of a time-periodic solution, one would have to choose appropriate boundary values. In the following, our main interest lies in the spatial decay of the velocity u and the associated vorticity $\operatorname{curl} u$.

A classical approach is to derive such properties with the help of fundamental solutions. For this purpose, we consider the problem in the whole space $\Omega = \mathbb{R}^3$, move the nonlinear terms to the right-hand side, and regard $(u, \mathfrak{p})$ as a solution to the *linear* time-periodic Oseen problem

$$\begin{cases} \partial_t u - \Delta u - \lambda \partial_1 u + \nabla \mathfrak{p} = \widetilde{f} & \text{in } \mathbb{T} \times \Omega, \\ \operatorname{div} u = 0 & \text{in } \mathbb{T} \times \Omega, \\ \lim_{|x| \to \infty} u(t, x) = 0 & \text{for } t \in \mathbb{T}, \end{cases}$$

with $\widetilde{f} := f - u \cdot \nabla u$. We then can express the solution by means of the time-periodic fundamental solutions introduced in Chapter 5 and obtain the identities

$$u(t, x) = \Gamma^\lambda * \widetilde{f}(t, x),$$
$$\operatorname{curl} u(t, x) = \int_{\mathbb{T} \times \mathbb{R}^3} \nabla \phi^\lambda(t - s, x - y) \wedge \widetilde{f}(s, y) \, \mathrm{d}(s, y).$$

The behavior of the velocity field of a time-periodic Navier–Stokes flow was already studied by GALDI and KYED [51], who established an asymptotic expansion for u. With the help of this expansion and the previous formula for u, we establish pointwise estimates of u and ∇u. By an iterative procedure, we improve the decay rates step by step until we arrive at an optimal rate that coincides with the decay of the corresponding fundamental solution Γ^λ and $\nabla \Gamma^\lambda$, respectively. As it turns out, this method cannot be applied directly to derive the optimal decay rate for $\operatorname{curl} u$. Instead, we follow an idea used by DEURING and GALDI [17] to regard the above formula as a fixed-point equation for the velocity field u. A suitable reformulation allows us to show existence of a fixed point z in a function class such that $\operatorname{curl} z$ has the expected decay properties and that $z(t, x) = u(t, x)$ for $|x|$ sufficiently large. Consequently, the decay rates of $\operatorname{curl} u$ and $\operatorname{curl} z$ coincide, which then concludes the proof.

One major discovery in Chapter 5 was that the properties of the steady-state part and the purely periodic part of the considered time-periodic fundamental solutions differ substantially. In view of this observation, we derive pointwise estimates for the steady-state part and the purely periodic part of u separately. In the end, we see that this difference in the decay rates also appears for both the velocity field and the vorticity field of a time-periodic Navier–Stokes flow, and that the respective purely periodic parts decay faster than the steady-state parts.

Observe that, from now on, when we give an estimate of the solution, then the constant in that estimate usually depends on the considered solution and thus on all other variables appearing in the system (6.1). For this reason, we now refrain from writing down the exact dependencies of the constants.

In Section 6.1 we derive pointwise estimates of u and ∇u. We express u by means of the time-periodic velocity fundamental solution and employ an iterative procedure to improve the decay estimates step by step. In Section 6.2 we investigate spatial decay of the vorticity $\operatorname{curl} u$. We express $\operatorname{curl} u$ via the time-periodic vorticity fundamental solution, and we derive decay estimates via a fixed-point argument. In Section 6.3 we apply our findings to Navier–Stokes flows in exterior domains.

6.1 The Velocity Field in the Whole Space

In the following, we consider the time-periodic Navier–Stokes equations in the three-dimensional whole space

$$\begin{cases} \partial_t u - \Delta u - \lambda \partial_1 u + u \cdot \nabla u + \nabla \mathfrak{p} = f & \text{in } \mathbb{T} \times \mathbb{R}^3, \\ \operatorname{div} u = 0 & \text{in } \mathbb{T} \times \mathbb{R}^3, \end{cases} \tag{6.2}$$

for a Reynolds number $\lambda > 0$ and a fixed time period $\mathcal{T} > 0$, which defines the torus group $\mathbb{T} = \mathbb{R}/\mathcal{T}\mathbb{Z}$. Throughout this chapter, we always consider weak solutions to (6.2) in the following sense.

Definition 6.1.1. Let $f \in \mathrm{L}^1_{\mathrm{loc}}(\mathbb{T} \times \mathbb{R}^3)^3$. A function $u \in \mathrm{L}^1_{\mathrm{loc}}(\mathbb{T} \times \mathbb{R}^3)^3$ is called *weak solution* to (6.2) if

i. $u \in \mathrm{L}^2(\mathbb{T}; \mathrm{D}^{1,2}_{0,\sigma}(\mathbb{R}^3))$,

ii. $\mathcal{P}_\perp u \in \mathrm{L}^\infty(\mathbb{T}; \mathrm{L}^2(\mathbb{R}^3))^3$,

iii. the identity

$$\int\limits_{\mathbb{T} \times \mathbb{R}^3} \left[-u \cdot \partial_t \varphi + \nabla u : \nabla \varphi - \lambda \partial_1 u \cdot \varphi + (u \cdot \nabla u) \cdot \varphi \right] \mathrm{d}(t,x) = \int\limits_{\mathbb{T} \times \mathbb{R}^3} f \cdot \varphi \, \mathrm{d}(t,x)$$

holds for all test functions $\varphi \in \mathrm{C}^\infty_{0,\sigma}(\mathbb{T} \times \mathbb{R}^3)$.

Remark 6.1.2. The existence of a weak solution with the above properties has been shown in [72, Theorem 6.3.1] for any $f \in L^2(\mathbb{T}; D_0^{-1,2}(\mathbb{R}^3))^3$. Therefore, this class seems to be a natural outset for further investigation. Nevertheless, at first glance, instead of ii. one would expect the condition $u \in L^\infty(\mathbb{T}; L^2(\mathbb{R}^3))^3$ instead, which naturally appears for weak solutions to the Navier–Stokes initial-value problem. From a physical perspective, this would mean that the flow has finite kinetic energy. However, this property cannot be expected for general time-periodic data f. As was shown by KYED [72, Theorem 5.2.4], for smooth data $f \in C_0^\infty(\mathbb{T} \times \mathbb{R}^3)^3$ one has $u \in L^\infty(\mathbb{T}; L^2(\mathbb{R}^3))^3$ if and only if $\int_{\mathbb{T} \times \mathbb{R}^3} f \, \mathrm{d}(x,t) = 0$. An analogous property was established by FINN [35] for the corresponding steady-state problem.

6.1.1 An Asymptotic Expansion

The asymptotic behavior of the solution to the time-periodic Navier–Stokes equations (6.2) was studied by GALDI and KYED in [51], who established an asymptotic expansion for the velocity field. Their main result reads as follows. For its statement, recall the steady-state Oseen fundamental solution $\Gamma_0^\lambda = \Gamma^O$ given in (5.9).

Theorem 6.1.3. *Let* $\lambda \neq 0$ *and* $f \in C_0^\infty(\mathbb{T} \times \mathbb{R}^3)^3$. *If* u *is a weak solution to* (6.2) *in the sense of Definition 6.1.1 such that*

$$\exists r \in (5, \infty): \quad \mathcal{P}_\perp u \in L^r(\mathbb{T} \times \mathbb{R}^3)^3, \tag{6.3}$$

then

$$u(t,x) = \Gamma_0^\lambda(x) \cdot \left(\int_{\mathbb{T} \times \mathbb{R}^3} f(t,x) \, \mathrm{d}(t,x) \right) + \mathcal{R}(t,x) \tag{6.4}$$

for all $(t, x) \in \mathbb{T} \times \mathbb{R}^3$, *where* $\mathcal{R}$ *satisfies*

$$\forall \varepsilon > 0 \, \exists C_{60} > 0 \, \forall |x| \geq 1, \, t \in \mathbb{T}: \quad |\mathcal{R}(t,x)| \leq C_{60} |x|^{-\frac{3}{2}+\varepsilon}. \tag{6.5}$$

Proof. See [51]. □

Remark 6.1.4. As pointed out in [51], the assumption (6.3) merely appears for technical reasons. It ensures additional local regularity, but it does not improve spatial decay of the solution as $|x| \to \infty$. For simplicity, one could thus always assume $(u, \mathfrak{p}) \in C^\infty(\mathbb{T} \times \mathbb{R}^3)^{3+1}$ instead of (6.3) and obtain the same results.

The leading term of the asymptotic expansion (6.4) is given by the steady-state Oseen fundamental solution Γ_0^λ. Further note that the above expansion is in accordance with the asymptotic expansion for the steady-state velocity field; see [42, Theorem X.8.2]. Note that the steady-state flow is a special case of time-periodic flow. Moreover, from Theorem 6.1.3 we immediately conclude the following decay estimates for the velocity field u.

Corollary 6.1.5. *Let u be as in Theorem 6.1.3. Then for all $\varepsilon > 0$ there exists $C_{61} > 0$ such that for all $|x| \geq 1$ and $t \in \mathbb{T}$ it holds*

$$|\mathcal{P}u(x)| \leq C_{61}|x|^{-1}(1 + s(\lambda x))^{-\frac{1}{2}+\varepsilon}, \qquad (6.6)$$

$$|\mathcal{P}_\perp u(t,x)| \leq C_{61}|x|^{-\frac{3}{2}+\varepsilon}, \qquad (6.7)$$

where $s(x) := |x| + x_1$.

Proof. From Theorem 6.1.3 we directly obtain

$$|\mathcal{P}u(x)| \leq c_0|\Gamma_0^\lambda(x)| + |\mathcal{P}\mathcal{R}(x)| \leq c_1\big([|x|(1 + s(\lambda x))]^{-1} + |x|^{-\frac{3}{2}+\varepsilon}\big),$$

where we used estimate (5.25) for $m = 0$. Now the elementary estimate $1 + s(\lambda x) \leq (1 + 2|\lambda|)|x|$ yields (6.6). Moreover, due to $\mathcal{P}_\perp \Gamma_0^\lambda = 0$, we have $\mathcal{P}_\perp u = \mathcal{P}_\perp \mathcal{R}$, which directly implies (6.7). $\qquad\square$

Remark 6.1.6. From the proof of Theorem 6.1.3 in [51], one can extract the much better decay estimate $|\mathcal{P}_\perp u(t,x)| \leq C_{62}|x|^{-\frac{12}{5}+\varepsilon}$ for the purely periodic part. However, the above result will be enough for our purposes, and we later derive an improved estimate anyway.

As we shall see in Theorem 6.1.9 below, the decay results of Corollary 6.1.5 are not optimal, which mainly has two reasons: Firstly, Theorem 6.1.3 does not treat the purely periodic part $\mathcal{P}_\perp u$ separately, so that its decay properties can merely be extracted from the remainder term $\mathcal{R}$ as we did in the previous proof. Secondly, the proof of Theorem 6.1.3 in [51] does not exploit the anisotropic decay estimate (5.25) and only makes use of the isotropic estimate $|\Gamma_0^\lambda(x)| \leq C_{63}|x|^{-1}$. If the argument in [51] had instead been based on estimate (5.25) in [51], then one could derive an improved (anisotropic) decay estimate of the remainder term $\mathcal{R}$.

Furthermore, from Theorem 6.1.9 we cannot extract information on the gradient ∇u of the velocity field, for which we also establish a decay result below. Besides being interesting in its own right, spatial estimates of ∇u are exploited in Section 6.2, where we derive decay properties of the vorticity field $\text{curl}\, u$.

6.1.2 Representation Formulas

Our approach for the derivation of pointwise estimates of the velocity field u is based on a representation via the time-periodic fundamental solution Γ^λ introduced in (5.37). More precisely, we utilize the identity $u = \Gamma^\lambda * (f - u \cdot \nabla u)$. In order to show this representation formula in a rigorous way, we employ the following regularity result.

Lemma 6.1.7. *Let $\lambda \neq 0$ and $f \in C_0^\infty(\mathbb{T} \times \mathbb{R}^3)^3$, and let u be a weak solution to (6.2) in the sense of Definition 6.1.1, which satisfies (6.3). Then $u \in C^\infty(\mathbb{T} \times \mathbb{R}^3)^3$ and*

$$\forall q \in (1,2),\, r \in (1,\infty): \quad \mathcal{P}u \in X_\lambda^q(\mathbb{R}^3) \cap D^{2,r}(\mathbb{R}^3)^3,$$
$$\forall q \in (1,\infty): \quad \mathcal{P}_\perp u \in W^{1,2,q}(\mathbb{T} \times \mathbb{R}^3)^3,$$

where $X_\lambda^q(\mathbb{R}^3)$ was introduced in (3.2). Moreover, there is a pressure function $\mathfrak{p} \in C^\infty(\mathbb{T} \times \mathbb{R}^3)$ such that (6.2) is satisfied pointwise and

$$\forall q \in (1,3),\, s \in (1,\infty): \quad \mathfrak{p} \in L^q(\mathbb{T}; L^{\frac{3q}{3-q}}(\Omega)) \cap L^s(\mathbb{T}; D^{1,s}(\Omega)).$$

Proof. This result was shown in [51, Lemma 5.1]. $\qquad\square$

Now we can derive the desired representation formula. We decompose it into formulas for the steady-state and the purely periodic part of the velocity, which are given by convolution with $\Gamma_0^\lambda = \Gamma^O$ and $\Gamma_\perp^\lambda$ defined in (5.9) and (5.39), respectively.

Proposition 6.1.8. *Let $\lambda \neq 0$ and $f \in C_0^\infty(\mathbb{T} \times \mathbb{R}^3)^3$, and let u be a weak solution to (6.2) in the sense of Definition 6.1.1, which satisfies (6.3). Then*

$$D_x^\alpha u = D_x^\alpha \Gamma^\lambda * [f - u \cdot \nabla u] \tag{6.8}$$

for all $\alpha \in \mathbb{N}_0^3$ with $|\alpha| \leq 1$. In particular, the steady-state part $v := \mathcal{P}u$ and the purely periodic part $w := \mathcal{P}_\perp u$ satisfy

$$D_x^\alpha v = D_x^\alpha \Gamma_0^\lambda * [\mathcal{P}f - v \cdot \nabla v - \mathcal{P}(w \cdot \nabla w)], \tag{6.9}$$
$$D_x^\alpha w = D_x^\alpha \Gamma_\perp^\lambda * [\mathcal{P}_\perp f - v \cdot \nabla w - w \cdot \nabla v - \mathcal{P}_\perp(w \cdot \nabla w)]. \tag{6.10}$$

Moreover, we have[1]

$$u = \Gamma^\lambda * f - \nabla \Gamma^\lambda * (u \otimes u) \tag{6.11}$$

---

[1]Here we set $(\nabla \Gamma^\lambda * U)_j := \sum_{\ell,m=1}^3 \partial_m \Gamma_{j\ell}^\lambda * U_{jm}$ for an $\mathbb{R}^{3\times3}$-valued function U.

and

$$v = \Gamma_0^\lambda * \mathcal{P}f - \nabla\Gamma_0^\lambda * \big[v \otimes v + \mathcal{P}(w \otimes w)\big], \tag{6.12}$$

$$w = \Gamma_\perp^\lambda * \mathcal{P}_\perp f - \nabla\Gamma_\perp^\lambda * \big[v \otimes w + w \otimes v + \mathcal{P}_\perp(w \otimes w)\big]. \tag{6.13}$$

Proof. From Lemma 6.1.7 we obtain $u \in L^r(\mathbb{T} \times \mathbb{R}^3)$ for all $r \in (2,\infty)$ and $\nabla u \in L^s(\mathbb{T} \times \mathbb{R}^3)$ for all $s \in (\frac{4}{3},\infty)$, so that $u \cdot \nabla u \in L^q(\mathbb{T} \times \mathbb{R}^3)$ for all $q \in (1,\infty)$. Since $\Gamma^\lambda = \Gamma_0^\lambda \otimes 1_\mathbb{T} + \Gamma_\perp^\lambda$, the function $U := \Gamma^\lambda * (f - u \cdot \nabla u)$ is well defined as a classical convolution integral by Theorem 5.1.8 and Theorem 5.2.3. With the same argument and the dominated convergence theorem, we further obtain $\partial_j U = \partial_j \Gamma^\lambda * (f - u \cdot \nabla u)$ for $j = 1,2,3$. Moreover, we have

$$\begin{aligned}
\mathcal{P}U &= (\Gamma_0^\lambda \otimes 1_\mathbb{T}) * \big[f - u \cdot \nabla u\big] = \Gamma_0^\lambda * \big[\mathcal{P}(f - u \cdot \nabla u)\big] \\
&= \Gamma_0^\lambda * \big[\mathcal{P}f - v \cdot \nabla v - \mathcal{P}(w \cdot \nabla w)\big], \\
\mathcal{P}_\perp U &= \Gamma_\perp^\lambda * \big[f - u \cdot \nabla u\big] = \Gamma_\perp^\lambda * \big[\mathcal{P}_\perp(f - u \cdot \nabla u)\big] \\
&= \Gamma_\perp^\lambda * \big[\mathcal{P}_\perp f - v \cdot \nabla w - w \cdot \nabla v - \mathcal{P}_\perp(w \cdot \nabla w)\big].
\end{aligned}$$

Therefore, (6.8), (6.9) and (6.10) follow if $U = u$. Since both U and u satisfy the time-periodic Oseen system (5.33) for suitable pressure functions $\mathfrak{p}$, the uniqueness statement from Lemma 5.2.5 implies $\mathcal{P}_\perp u = \mathcal{P}_\perp U$ and that $\mathcal{P}u - \mathcal{P}U$ is a polynomial in each component. With Young's inequality we obtain

$$\|\mathcal{P}U\|_6 \le \|\Gamma_0^\lambda\|_{12/5}\|\mathcal{P}(f - u \cdot \nabla u)\|_{12/9} < \infty$$

since $\Gamma_0^\lambda \in L^{12/5}(\mathbb{R}^3)$ by Theorem 5.1.8. Therefore, $\mathcal{P}u - \mathcal{P}U \in L^6(\mathbb{R}^3)$, which implies $\mathcal{P}u = \mathcal{P}U$. In total, we thus have $u = U = \Gamma^\lambda * (f - u \cdot \nabla u)$. The remaining formulas (6.11), (6.12) and (6.13) now follow from the identity $u \cdot \nabla u = \text{div}(u \otimes u)$ due to $\text{div}\, u = 0$. $\qquad\square$

6.1.3 Spatial Decay Estimates

Now we employ the representation formulas (6.8) and (6.11) in order to derive a decay estimate of u and ∇u. The proof is based on an iterative procedure: For u we already have a pointwise estimate by Corollary 6.1.5. This immediately yields a (faster) decay rate of $u \otimes u$, which we can use to obtain a new estimate of u via (6.11). Repeating this argument, we can improve the estimate of u iteratively. These new estimates lead to an even better estimate of $u \otimes u$. Employing (6.11) again, we then deduce another estimate of u. Note that, since $f \in C_0^\infty(\mathbb{T} \times \mathbb{R}^3)$, we directly conclude an estimate of the term $\Gamma^\lambda * f$ from Theorem 5.1.7 and Theorem 5.2.8. Since this term appears in each step of the above iteration, we cannot improve the decay estimate of u beyond that of $\Gamma^\lambda * f$, which coincides with the decay rate of Γ^λ.

The proof for the decay rate of ∇u works slightly different. In this case, we do not have a decay rate initially, which is why we use integrability results from Lemma 6.1.7 in order to deduce a first spatial estimate of ∇u from the representation (6.8). While this will not be optimal, we can then employ the above iteration scheme again until we arrive at the decay rate of $\nabla(\Gamma^\lambda * f)$, which coincides with that of $\nabla \Gamma^\lambda$.

As the purely periodic part decays faster than the steady-state part of the fundamental solution (compare Theorem 5.1.7 with Theorem 5.2.8), one can also expect that the purely periodic part of the velocity field decays faster than its steady-state part. We therefore establish separate spatial decay estimates of both parts of u and ∇u by exploiting the decomposed representation formulas from Proposition 6.1.8.

Note that the constant appearing in the estimates for u depends on the solution u itself.

Theorem 6.1.9. *Let $\lambda \neq 0$ and $f \in C_0^\infty(\mathbb{T} \times \mathbb{R}^3)^3$, and let u be a weak time-periodic solution to (6.2) in the sense of Definition 6.1.1, which satisfies (6.3). Then there is $C_{64} > 0$ such that for all $t \in \mathbb{T}$ and $x \in \mathbb{R}^3$ the function u satisfies*

$$|\mathcal{P}u(x)| \leq C_{64}\big[(1 + |x|)(1 + s(\lambda x))\big]^{-1}, \tag{6.14}$$

$$|\nabla\mathcal{P}u(x)| \leq C_{64}\big[(1 + |x|)(1 + s(\lambda x))\big]^{-\frac{3}{2}}, \tag{6.15}$$

$$|\mathcal{P}_\perp u(t, x)| \leq C_{64}(1 + |x|)^{-3}, \tag{6.16}$$

$$|\nabla\mathcal{P}_\perp u(t, x)| \leq C_{64}(1 + |x|)^{-4}. \tag{6.17}$$

Proof. We split $u = v + w$ into a steady-state part $v \coloneqq \mathcal{P}u$ and a purely periodic part $w \coloneqq \mathcal{P}_\perp u$. By Lemma 6.1.7 we have $u \in C^\infty(\mathbb{T} \times \mathbb{R}^3)^3$, and thus

$$|v(x)| + |\nabla v(x)| + |w(t, x)| + |\nabla w(t, x)| \leq c_0 \tag{6.18}$$

for all $t \in \mathbb{T}$ and $|x| \leq 1$. Combining (6.18) with Corollary 6.1.5 (with $\varepsilon = 1/4$), we conclude

$$|v(x)| \leq c_1(1 + |x|)^{-1}(1 + s(\lambda x))^{-1/4}, \tag{6.19}$$

$$|w(t, x)| \leq c_2(1 + |x|)^{-5/4} \tag{6.20}$$

for all $t \in \mathbb{T}$ and $x \in \mathbb{R}^3$. This implies

$$\big|v \otimes v + \mathcal{P}[w \otimes w]\big|(x) \leq c_3\big((1 + |x|)^{-2}(1 + s(\lambda x))^{-1/2} + (1 + |x|)^{-5/2}\big)$$
$$\leq c_4(1 + |x|)^{-2}(1 + s(\lambda x))^{-1/2},$$

and the representation formula (6.12) in combination with $\mathcal{P}f \in C_0^\infty(\mathbb{R}^3)$ and Theorem 5.1.7 yields

$$|v(x)| \leq \big|\Gamma_0^\lambda * \mathcal{P}f\big|(x) + \big|\nabla\Gamma_0^\lambda * \big[v \otimes v + \mathcal{P}[w \otimes w]\big]\big|(x)$$
$$\leq c_5\big(\big[(1 + |x|)(1 + s(\lambda x))\big]^{-1} + (1 + |x|)^{-5/4}\big(1 + s(\lambda x)\big)^{-3/4}\big)$$
$$\leq c_6\big[(1 + |x|)(1 + s(\lambda x))\big]^{-1},$$

which is the desired estimate (6.14).

Now (6.14) together with (6.20) leads to

$$\big|v \otimes w + w \otimes v + \mathcal{P}_\perp[w \otimes w]\big|(t, x) \leq c_7\big((1 + |x|)^{-9/4} + (1 + |x|)^{-5/2}\big) \qquad (6.21)$$
$$\leq c_8(1 + |x|)^{-9/4}.$$

Therefore, the representation formula (6.13) in combination with $\mathcal{P}_\perp f \in C_0^\infty(\mathbb{T} \times \mathbb{R}^3)$ and Theorem 5.2.8 implies

$$|w(t, x)| \leq \big|\Gamma_\perp^\lambda * \mathcal{P}_\perp f\big|(t, x) + \big|\nabla\Gamma_\perp^\lambda * \big[v \otimes w + w \otimes v + \mathcal{P}_\perp[w \otimes w]\big]\big|(t, x)$$
$$\leq c_9\big((1 + |x|)^{-3} + (1 + |x|)^{-9/4}\big) \leq c_{10}(1 + |x|)^{-9/4}.$$

Using this estimate and (6.14) again, we conclude

$$\big|v \otimes w + w \otimes v + \mathcal{P}_\perp[w \otimes w]\big|(t, x) \leq c_{11}\big((1 + |x|)^{-13/4} + (1 + |x|)^{-9/2}\big) \qquad (6.22)$$
$$\leq c_{12}(1 + |x|)^{-13/4}.$$

Repeating the above argument with (6.22) instead of (6.21), we end up with (6.16).

Now let us turn to the estimates of ∇u. Due to (6.18), the estimates (6.15) and (6.17) hold for all $t \in \mathbb{T}$ and $|x| \leq 2$, and it suffices to consider $|x| \geq 2$ in the following. Let $R := |x|/2 \geq 1$. By Proposition 6.1.8, for $j \in \{1, 2, 3\}$ we have

$$\partial_j v(x) = I_1(x) + I_2(x) + I_3(x),$$
$$\partial_j w(t, x) = J_1(t, x) + J_2(t, x) + J_3(t, x) + J_4(t, x)$$

with

$$I_1 := \partial_j \Gamma_0^\lambda * \mathcal{P}f, \qquad\qquad J_1 := \partial_j \Gamma_\perp^\lambda * \mathcal{P}_\perp f,$$
$$I_2 := \partial_j \Gamma_0^\lambda * \left[-v \cdot \nabla v \right], \qquad J_2 := \partial_j \Gamma_\perp^\lambda * \left[-v \cdot \nabla w \right],$$
$$I_3 := \partial_j \Gamma_0^\lambda * \left[-\mathcal{P}[w \cdot \nabla w] \right], \qquad J_3 := \partial_j \Gamma_\perp^\lambda * \left[-w \cdot \nabla v \right],$$
$$J_4 := \partial_j \Gamma_\perp^\lambda * \left[-\mathcal{P}_\perp[w \cdot \nabla w] \right].$$

We estimate these terms separately. Since $f \in C_0^\infty(\mathbb{R}^3)$, an application of Theorem 5.1.7 leads to

$$|I_1(x)| \le c_{13} \left[(1 + |x|)(1 + s(\lambda x)) \right]^{-3/2}. \tag{6.23}$$

We decompose I_2 and estimate $|I_2| \le I_{21} + I_{22}$ with

$$I_{21}(x) := \int_{B_R} |\partial_j \Gamma_0^\lambda(x - y)| \|v(y)\| |\nabla v(y)| \, \mathrm{d}y,$$

$$I_{22}(x) := \int_{B^R} |\partial_j \Gamma_0^\lambda(x - y)| \|v(y)\| |\nabla v(y)| \, \mathrm{d}y.$$

Since $|y| \le R$ implies $|x - y| \ge |x|/2 = R \ge 1$, the pointwise estimate (5.25) yields

$$I_{21}(x) \le \int_{B_R} \left[(1 + |x - y|)(1 + s(x - y)) \right]^{-3/2} |v(y)| |\nabla v(y)| \, \mathrm{d}y$$
$$\le c_{14}(1 + |x|)^{-3/2} \|v\|_3 \|\nabla v\|_{\frac{3}{2}} \le c_{15}(1 + |x|)^{-3/2}$$

because $v \in L^3(\mathbb{R}^3)$ and $\nabla v \in L^{3/2}(\mathbb{R}^3)$ by Lemma 6.1.7. Moreover, we have $\partial_j \Gamma_0^\lambda \in L^{17/12}(\mathbb{R}^3)$ and $\nabla v \in L^{17/5}(\mathbb{R}^3)$ due to estimates (5.28) and (5.29) and Lemma 6.1.7, which leads to

$$I_{22}(x) \le c_{16} \|\partial_j \Gamma_0^\lambda\|_{\frac{17}{12}} \|\nabla v\|_{\frac{17}{5}} \|v\|_{L^\infty(B^R)} \le c_{17}(1 + |x|)^{-1}$$

by (6.14). For I_3 we proceed similarly. We estimate $|I_3| \le I_{31} + I_{32}$, where

$$I_{31}(x) := \int_{B_R} |\partial_j \Gamma_0^\lambda(x - y)| \|\mathcal{P}(w \cdot \nabla w)(y)\| \, \mathrm{d}y,$$

$$I_{32}(x) := \int_{B^R} |\partial_j \Gamma_0^\lambda(x - y)| \|\mathcal{P}(w \cdot \nabla w)(y)\| \, \mathrm{d}y.$$

Arguing as for I_{21} and I_{22}, we obtain

$$I_{31}(x) \leq c_{18} \int_{B_R} \left[(1 + |x - y|)(1 + s(x - y)) \right]^{-3/2} \int_{\mathbb{T}} |w(t, y)| |\nabla w(t, y)| \, dt dy$$

$$\leq c_{19}(1 + |x|)^{-3/2} \|w\|_{\frac{3}{2}} \|\nabla w\|_3 \leq c_{20}(1 + |x|)^{-3/2}$$

and

$$I_{32}(x) \leq c_{21} \|\partial_j \Gamma_0^\lambda\|_{\frac{17}{12}} \|\nabla w\|_{L^1(\mathbb{T}; L^{\frac{17}{5}}(\mathbb{R}^3))} \|w\|_{L^\infty(\mathbb{T} \times B^R)} \leq c_{22}(1 + |x|)^{-3}.$$

Collecting the above estimates, we thus conclude

$$|\nabla v(x)| \leq c_{23}(1 + |x|)^{-1} \tag{6.24}$$

for $|x| \geq 2$. By (6.18), this estimates also holds for $|x| \leq 2$ and thus for all $x \in \mathbb{R}^3$. In order to improve the estimate, we next consider $\partial_j w$. From $f \in C_0^\infty(\mathbb{T} \times \mathbb{R}^3)$ and Theorem 5.2.8, we directly deduce

$$|J_1(t, x)| \leq c_{24}(1 + |x|)^{-4}. \tag{6.25}$$

We decompose J_2 and estimate $|J_2| \leq J_{21} + J_{22}$ with

$$J_{21}(t, x) := \int_{\mathbb{T}} \int_{B_R} |\partial_j \Gamma_\perp^\lambda(t - s, x - y)| |v(y)| |\nabla w(s, y)| \, dy ds,$$

$$J_{22}(t, x) := \int_{\mathbb{T}} \int_{B^R} |\partial_j \Gamma_\perp^\lambda(t - s, x - y)| |v(y)| |\nabla w(s, y)| \, dy ds.$$

By Hölder's inequality in space and time, from (5.41) we obtain

$$J_{21}(t, x) \leq \left(\int_{B_R} \int_{\mathbb{T}} |\partial_j \Gamma_\perp^\lambda(t - s, x - y)|^2 \, ds dy \right)^{\frac{1}{2}} \|v\|_4 \|\nabla w\|_4$$

$$\leq c_{25} \left(\int_{B_R} |x - y|^{-8} \, dy \right)^{\frac{1}{2}} \|v\|_4 \|\nabla w\|_4$$

$$\leq c_{26} |x|^{-4} R^{\frac{3}{2}} \|v\|_4 \|\nabla w\|_4 \leq c_{27} |x|^{-5/2}$$

since $v \in L^4(\mathbb{R}^3)$ and $\nabla w \in L^4(\mathbb{T} \times \mathbb{R}^3)$ by Lemma 6.1.7. Another application of Hölder's inequality and the decay estimate (6.14) yield

$$J_{22}(t, x) \leq \int_{\mathbb{T}} \int_{B^R} |\partial_j \Gamma_\perp^\lambda(t - s, x - y)| \, dy ds \sup_{(s, y) \in \mathbb{T} \times B^R} |v(y) \cdot \nabla w(s, y)|$$

$$\leq c_{28} \|\partial_j \Gamma_\perp^\lambda\|_1 \|v\|_{L^\infty(B^R)} \|\nabla w\|_\infty \leq c_{29} |x|^{-1}$$

because $\nabla \Gamma_\perp^\lambda \in L^1(\mathbb{T} \times \mathbb{R}^3)$ by (5.44) and $\nabla w \in L^\infty(\mathbb{T} \times \mathbb{R}^3)$ by Lemma 6.1.7 and Sobolev embeddings. In a similar fashion, we can use (6.16) to estimate J_3 and J_4 and obtain

$$|J_3(t,x)| \le c_{30}\left(|x|^{-\frac{5}{2}}\|w\|_4\|\nabla v\|_4 + |x|^{-3}\|\partial_j\Gamma_\perp^\lambda\|_{\frac{9}{8}}\|\nabla v\|_9\right) \le c_{31}|x|^{-\frac{5}{2}},$$

$$|J_4(t,x)| \le c_{32}\left(|x|^{-\frac{5}{2}}\|w\|_4\|\nabla w\|_4 + |x|^{-3}\|\partial_j\Gamma_\perp^\lambda\|_1\|\nabla w\|_\infty\right) \le c_{33}|x|^{-\frac{5}{2}}.$$

Collecting the above estimates, we end up with

$$|\nabla w(t,x)| \le c_{34}(1+|x|)^{-1} \tag{6.26}$$

for $|x| \ge 2$. Again, (6.18) implies that this estimate also holds for $|x| \le 2$, and thus for all $(t,x) \in \mathbb{T} \times \mathbb{R}^3$.

With (6.26) at hand, we can now improve (6.24) by a bootstrap argument. By (6.14), (6.24), (6.16) and (6.26), we have

$$\begin{aligned}
&\left|v(x) \cdot \nabla v(x) + \mathcal{P}[w \cdot \nabla w](x)\right| \\
&\quad \le c_{35}\left((1+|x|)^{-2}(1+s(\lambda x))^{-1} + (1+|x|)^{-4}\right) \\
&\quad \le c_{36}(1+|x|)^{-2}(1+s(\lambda x))^{-1/2},
\end{aligned}$$

so that Theorem 5.1.7 implies

$$|I_2(x) + I_3(x)| \le c_{37}(1+|x|)^{-5/4}(1+s(\lambda x))^{-3/4}.$$

Together with (6.23) we thus obtain

$$|\nabla v(x)| \le c_{38}(1+|x|)^{-5/4}(1+s(\lambda x))^{-3/4},$$

so that from (6.14), (6.16) and (6.26) we deduce

$$\begin{aligned}
&\left|v(x) \cdot \nabla v(x) + \mathcal{P}[w \cdot \nabla w](x)\right| \\
&\quad \le c_{39}\left((1+|x|)^{-9/4}(1+s(\lambda x))^{-7/4} + (1+|x|)^{-4}\right) \\
&\quad \le c_{40}(1+|x|)^{-9/4}(1+s(\lambda x))^{-7/4}.
\end{aligned}$$

Another application of Theorem 5.1.7 now leads to

$$|I_2(x) + I_3(x)| \le c_{41}\left[(1+|x|)(1+s(\lambda x))\right]^{-3/2},$$

and by a combination with (6.23) we arrive at (6.15).

For the derivation of (6.17) we employ a similar bootstrap argument. Now from (6.14), (6.15), (6.16) and (6.26) we deduce

$$N(t,x) := \left|v(x)\cdot\nabla w(t,x) + w(t,x)\cdot\nabla v(x) + \mathcal{P}_\perp\big[w(t,x)\cdot\nabla w(t,x)\big]\right|$$
$$\leq c_{42}\big((1+|x|)^{-2} + (1+|x|)^{-\frac{9}{2}} + (1+|x|)^{-4}\big) \leq c_{43}(1+|x|)^{-2},$$

so that Theorem 5.2.8 implies

$$J(t,x) := |J_2(t,x) + J_3(t,x) + J_4(t,x)| \leq c_{44}(1+|x|)^{-2}.$$

Combining this with (6.25), we conclude

$$|\nabla w(t,x)| \leq c_{45}(1+|x|)^{-2}, \tag{6.27}$$

which we use together with (6.14), (6.15) and (6.16) to deduce

$$N(t,x) \leq c_{46}(1+|x|)^{-3} \leq c_{47}(1+|x|)^{-5/2}.$$

This results in $J(t,x) \leq c_{48}(1+|x|)^{-5/2}$ by Theorem 5.2.8, and from (6.25) we conclude

$$|\nabla w(t,x)| \leq c_{49}(1+|x|)^{-5/2}.$$

A combination of this estimate with (6.14), (6.15), (6.16) leads to $N(t,x) \leq c_{50}(1+|x|)^{-7/2}$. Hence, $J(t,x) \leq c_{51}(1+|x|)^{-7/2}$ by Theorem 5.2.8, and together with (6.25) we have

$$|\nabla w(t,x)| \leq c_{52}(1+|x|)^{-7/2}.$$

In combination with (6.14), (6.15) and (6.16), this implies the estimate $N(t,x) \leq c_{53}(1+|x|)^{-9/2}$, and thus $J(t,x) \leq c_{54}(1+|x|)^{-4}$. Exploiting (6.25) again, we finally conclude the remaining estimate (6.17). $\qquad\square$

6.2 The Vorticity Field in the Whole Space

After the deduction of spatial decay estimates of u and ∇u in Theorem 6.1.9, this section is dedicated to the deduction of analogous estimates of the vorticity $\operatorname{curl} u$. In virtue of the proof of Theorem 6.1.9, it seems natural to exploit properties of the vorticity fundamental solution ϕ^λ introduced in (5.56).

By Theorem 6.1.9 we already have a decay estimate for the vorticiy $\operatorname{curl} u$, namely

$$|\mathcal{P}\operatorname{curl} u(x)| \leq C_{65}|\nabla\mathcal{P}u(x)| \leq C_{66}\big[(1+|x|)\big(1+s(\lambda x)\big)\big]^{-\frac{3}{2}},$$
$$|\mathcal{P}_\perp\operatorname{curl} u(t,x)| \leq C_{67}|\nabla\mathcal{P}_\perp u(t,x)| \leq C_{68}\big(1+|x|\big)^{-4}.$$

However, comparing these decay rates to those of the vorticity fundamental solution $\phi^\lambda = \phi_0^\lambda \otimes 1_\mathbb{T} + \phi_\perp^\lambda$ given in Theorem 5.3.1 and Theorem 5.3.3, they seem not to be optimal since one would also expect some kind of exponential decay (at least outside of the wake region). Note that steady state solutions (6.2) show this behavior; see [14, 6]. Indeed, this is the case as we shall see in the following.

Starting from the above decay rates for $\operatorname{curl} u$, one could think of proceeding as in the proof of Theorem 6.1.9 and successively increasing the decay rate of $\operatorname{curl} u$ with the help of an appropriate representation formula by a bootstrap argument. Indeed, an according formula is available as we shall see in Section 6.2.1. However, starting from the above *polynomial* decay rate, this iteration would only result in polynomial decay rates and thus cannot capture the expected decay of exponential type, which is suggested by the decay estimates of ϕ^λ. Therefore, we employ a different approach: We express u as a fixed point of a certain mapping F_S and show that the fixed-point equation $z = F_S(z)$ has a solution in a certain class of functions z such that $\operatorname{curl} z$ has the expected exponential decay rate. Afterwards, we show uniqueness of the solution to this fixed-point equation, but in the larger class of functions that merely share the decay rate of u established in Theorem 6.1.9. This implies $u = z$ and results in the desired estimates for $\operatorname{curl} u$.

Since the representation formula we derive for $\operatorname{curl} u$ is of the form (5.54), the final decay rate will be governed by that of $\nabla \phi^\lambda$. However, while the steady-state part of $\operatorname{curl} u$ has the same decay rate as the steady-state part $\nabla \phi_0^\lambda$, the purely periodic part of $\operatorname{curl} u$ decays slower than $\nabla \phi_\perp^\lambda$ and also exhibits a wake-like behavior. This phenomenon is due to the interaction of the nonlinear terms and is further explained in Remark 6.2.2 below.

6.2.1 Representation Formulas

In order to derive a decay estimate for the vorticity $\operatorname{curl} u$ from the properties of the vorticity fundamental solution ϕ^λ, we need to connect these objects in an appropriate way as we already did in Subsection 6.1.2 for the velocity. A simple application of curl to both sides of (6.8) and a computation as at the beginning of Subsection 5.3.1 would lead to the formula

$$\operatorname{curl} u(t, x) = \int_G \nabla \phi^\lambda(t - s, x - y) \wedge \left[f - u \cdot \nabla u \right](s, y) \, \mathrm{d}(s, y),$$

where we recall the abbreviation $G := \mathbb{T} \times \mathbb{R}^3$. However, this representation cannot yield decay of exponential type since Theorem 6.1.9 merely gives a polynomial decay rate of $u \cdot \nabla u$. Therefore, we need different representation formulas, which we introduce in the next proposition.

Proposition 6.2.1. *Let $\lambda \neq 0$ and $f \in C_0^\infty(\mathbb{T} \times \mathbb{R}^3)^3$, and let u be a weak solution to (6.2) in the sense of Definition 6.1.1, which satisfies (6.3). Then*

$$D_x^\alpha u = D_x^\alpha \Gamma^\lambda * [f - \operatorname{curl} u \wedge u] \tag{6.28}$$

for all $\alpha \in \mathbb{N}_0^3$ with $|\alpha| \leq 1$. In particular, the steady-state part $v := \mathcal{P}u$ and the purely periodic part $w := \mathcal{P}_\perp u$ satisfy

$$D_x^\alpha v = D_x^\alpha \Gamma_0^\lambda * \left[\mathcal{P}f - \operatorname{curl} v \wedge v - \mathcal{P}(\operatorname{curl} w \wedge w)\right], \tag{6.29}$$

$$D_x^\alpha w = D_x^\alpha \Gamma_\perp^\lambda * \left[\mathcal{P}_\perp f - \operatorname{curl} v \wedge w - \operatorname{curl} w \wedge v + \mathcal{P}_\perp(\operatorname{curl} w \wedge w)\right]. \tag{6.30}$$

Moreover, we have

$$\operatorname{curl} u(t, x) = \int_G \nabla \phi^\lambda(t - s, x - y) \wedge [f - \operatorname{curl} u \wedge u](s, y) \, \mathrm{d}(s, y), \tag{6.31}$$

and

$$\operatorname{curl} v(x) = \int_{\mathbb{R}^3} \nabla \phi_0^\lambda(x - y) \wedge \left[\mathcal{P}f - \operatorname{curl} v \wedge v - \mathcal{P}(\operatorname{curl} w \wedge w)\right](y) \, \mathrm{d}y, \tag{6.32}$$

as well as

$$\operatorname{curl} w(t, x) = \int_G \nabla \phi_\perp^\lambda(t - s, x - y) \wedge \left[\mathcal{P}_\perp f - \operatorname{curl} v \wedge w \right. \tag{6.33}$$
$$\left. - \operatorname{curl} w \wedge v - \mathcal{P}_\perp(\operatorname{curl} w \wedge w)\right](s, y) \, \mathrm{d}(s, y).$$

Proof. A straightforward calculation shows that

$$u \cdot \nabla u = \frac{1}{2} \nabla(|u|^2) + \operatorname{curl} u \wedge u.$$

Since $\Gamma^\lambda * \nabla(|u|^2) = \operatorname{div}(\Gamma^\lambda * |u|^2) = 0$ due to $(5.35)_2$, the equations (6.28), (6.29) and (6.30) are direct consequences of (6.8). Repeating now the computations from the beginning of Subsection 5.3.1 with f replaced with $f - \operatorname{curl} u \wedge u$, we conclude the remaining identities. $\qquad \square$

Remark 6.2.2. Let us have a closer look at the formulas (6.32) and (6.33) and the decay rates we can expect to derive for $\operatorname{curl} v$ and $\operatorname{curl} w$. As we see below, $\operatorname{curl} v$ obeys the estimate

$$|\operatorname{curl} v(x)| \le C_{69}|x|^{-3/2}\, e^{-C_{70}s(\lambda x)} \tag{6.34}$$

for $|x|$ sufficiently large. This means that the decay rate of the steady-state part $\operatorname{curl} v$ of the vorticity coincides with the associated integral kernel $\nabla\phi_0^\lambda$, which is the best we can expect for general $f \in C_0^\infty(G)$. This is the same phenomenon we discovered for the steady-state and purely periodic parts of u and ∇u. In contrast, we find

$$|\operatorname{curl} w(t, x)| \le C_{71}|x|^{-9/2}\, e^{-C_{72}s(\lambda x)}$$

for $|x|$ sufficiently large. This decay is not as fast as that of $\nabla\phi_1^\lambda$, which decays exponentially in every direction; see (5.66). This discrepancy is due to the term $\operatorname{curl} v \wedge w$ appearing in (6.33). Assuming the optimal decay rates of $\operatorname{curl} v$ above and of w from (6.17), we conclude the estimate

$$|\operatorname{curl} v \wedge w|(t, x) \le C_{73}|x|^{-9/2}\, e^{-C_{74}s(\lambda x)},$$

which cannot be improved. Therefore, this slower decay rate dominates the spatial decay estimates of $\operatorname{curl} w$. Note that although $\operatorname{curl} w$ does not decay as fast as $\nabla\phi_1^\lambda$, its decay is faster than that of the steady-state part $\operatorname{curl} v$.

6.2.2 A Decomposition of the Velocity Field

Here we derive the fixed-point equation our approach is based on. For this purpose, we express u by means of the representation (6.28), which we decompose into the sum of two terms.

Let $\chi \in C_0^\infty(\mathbb{R}; [0, 1])$ with $\chi(s) = 1$ for $|s| \le 5/4$ and $\chi(s) = 0$ for $|s| \ge 7/4$. For $S > 0$ define the function $\chi_S \in C_0^\infty(\mathbb{R}^3; [0, 1])$ by $\chi_S(x) := \chi(S^{-1}|x|)$. Consider $S_0 > 0$ such that $\operatorname{supp} f \subset \mathbb{T} \times B_{S_0}$. For $S \in [2S_0, \infty)$ we express the representation formula (6.28) as the sum of two terms, namely

$$u = \Gamma^\lambda * \left[-(1 - \chi_S)\operatorname{curl} u \wedge u \right] + \Gamma^\lambda * \left[f - \chi_S \operatorname{curl} u \wedge u \right].$$

Due to $\operatorname{supp}(1 - \chi_S) \subset B^S$, this yields

$$u|_{\mathbb{T}\times B^S} = \mathcal{F}_S(u|_{\mathbb{T}\times B^S}) + \mathcal{H}_S, \tag{6.35}$$

where

$$\mathcal{F}_S(z) := \big(\Gamma^\lambda * \big[-(1-\chi_S)\operatorname{curl} z \wedge z\big]\big)\big|_{\mathbb{T}\times B^S},$$
$$\mathcal{H}_S := \big(\Gamma^\lambda * \big[f - \chi_S \operatorname{curl} u \wedge u\big]\big)\big|_{\mathbb{T}\times B^S}.$$

We set

$$\mathcal{A}(z) := -\operatorname{curl} z \wedge z.$$

Then, with $z_0 := \mathcal{P}z$ and $z_\perp := \mathcal{P}_\perp z$ we have

$$\mathcal{A}_0(z) := \mathcal{P}\mathcal{A}(z) = -\operatorname{curl} z_0 \wedge z_0 - \mathcal{P}(\operatorname{curl} z_\perp \wedge z_\perp), \tag{6.36}$$
$$\mathcal{A}_\perp(z) := \mathcal{P}_\perp\mathcal{A}(z) = -\operatorname{curl} z_0 \wedge z_\perp - \operatorname{curl} z_\perp \wedge z_0 - \mathcal{P}_\perp(\operatorname{curl} z_\perp \wedge z_\perp), \tag{6.37}$$

and for $\alpha \in \mathbb{N}_0^3$ with $|\alpha| \le 1$ we obtain

$$D_x^\alpha \mathcal{P}\mathcal{F}_S(z)(x) = \big(D_x^\alpha \Gamma_0^\lambda * \big[(1-\chi_S)\mathcal{A}_0(z)\big]\big)(x), \tag{6.38}$$
$$D_x^\alpha \mathcal{P}_\perp \mathcal{F}_S(z)(t,x) = \big(D_x^\alpha \Gamma_\perp^\lambda * \big[(1-\chi_S)\mathcal{A}_\perp(z)\big]\big)(t,x), \tag{6.39}$$
$$\operatorname{curl}\mathcal{P}\mathcal{F}_S(z)(x) = \int_{\mathbb{R}^3} \nabla\phi_0^\lambda(x-y) \wedge \big[(1-\chi_S)\mathcal{A}_0(z)\big](y)\,\mathrm{d}y, \tag{6.40}$$

$$\operatorname{curl}\mathcal{P}_\perp\mathcal{F}_S(z)(t,x)$$
$$= \int_{\mathbb{T}\times\mathbb{R}^3} \nabla\phi_\perp^\lambda(t-s,x-y) \wedge \big[(1-\chi_S)\mathcal{A}_\perp(z)\big](s,y)\,\mathrm{d}(s,y), \tag{6.41}$$

and

$$D_x^\alpha \mathcal{P}\mathcal{H}_S(x) = \big(D_x^\alpha \Gamma_0^\lambda * \big[\mathcal{P}f + \chi_S\mathcal{A}_0(u)\big]\big)(x), \tag{6.42}$$
$$D_x^\alpha \mathcal{P}_\perp \mathcal{H}_S(t,x) = \big(D_x^\alpha \Gamma_\perp^\lambda * \big[\mathcal{P}_\perp f + \chi_S\mathcal{A}_\perp(u)\big]\big)(t,x), \tag{6.43}$$
$$\operatorname{curl}\mathcal{P}\mathcal{H}_S(x) = \int_{\mathbb{R}^3} \nabla\phi_0^\lambda(x-y) \wedge \big[\mathcal{P}f + \chi_S\mathcal{A}_0(u)\big](y)\,\mathrm{d}y, \tag{6.44}$$

$$\operatorname{curl}\mathcal{P}_\perp\mathcal{H}_S(t,x)$$
$$= \int_{\mathbb{T}\times\mathbb{R}^3} \nabla\phi_\perp^\lambda(t-s,x-y) \wedge \big[\mathcal{P}_\perp f + \chi_S\mathcal{A}_\perp(u)\big](s,y)\,\mathrm{d}(s,y) \tag{6.45}$$

for all $t \in \mathbb{T}$ and $x \in \mathbb{R}^3$ with $|x| > S$.

6.2.3 Function Spaces

We introduce the functional framework. Let $\varepsilon \in (0,\tfrac{1}{4})$ and fix a radius $S > 0$. We define the following (semi-)norms, which take into account the different decay rates of the steady-state and the purely periodic parts:

$$M_S(z) := \operatorname*{ess\,sup}_{x \in B^S} \left[|x|(1 + s(x)) |\mathcal{P}z(x)| + \left[|x|(1 + s(x)) \right]^{3/2} |\nabla \mathcal{P}z(x)| \right]$$

$$+ \operatorname*{ess\,sup}_{(t,x) \in \mathbb{T} \times B^S} \left[|x|^3 |\mathcal{P}_\perp z(t,x)| + |x|^4 |\nabla \mathcal{P}_\perp z(t,x)| \right],$$

$$N_S^\varepsilon(z) := \operatorname*{ess\,sup}_{x \in B^S} |x|^{3/2} \, e^{\frac{s(Kx)}{1+S}} \, |\operatorname{curl} \mathcal{P}z(x)|$$

$$+ \operatorname*{ess\,sup}_{(t,x) \in \mathbb{T} \times B^S} |x|^{9/2 - \varepsilon} \, e^{\frac{s(Kx)}{1+S}} \, |\operatorname{curl} \mathcal{P}_\perp z(t,x)|,$$

where $K := \frac{1}{4} \operatorname{sgn}(\lambda) \min\{|\lambda|, C_{58}\}$ with C_{58} from Theorem 5.3.3.

We defined $M_S(z)$ in the above way in order to adequately capture the asymptotic behavior of u and ∇u. Recall that we have $M_S(u|_{\mathbb{T} \times B^S}) < \infty$ due to Theorem 6.1.9.

In contrast, the motivation for the definition of $N_S^\varepsilon(z)$ requires more explanation. First, let us focus on the exponential term. How the denominator $1 + S$ comes into play will become clear in Lemma 6.2.4 below. To explain the choice of the constant K, note that for $K > 0$ we have $\lambda > 0$ and $\lambda \geq 4K$, so that

$$s(Kx) = K(|x| + x_1) \leq \frac{1}{4}\lambda(|x| + x_1) = \frac{1}{4}s(\lambda x),$$

and for $K < 0$ we have $\lambda < 0$ and $-\lambda \geq -4K$, so that

$$s(Kx) = -K(|x| - x_1) \leq -\frac{1}{4}\lambda(|x| - x_1) = \frac{1}{4}s(\lambda x).$$

In total, we thus see

$$s(Kx) \leq \frac{1}{4}s(\lambda x).$$

Additionally, we have

$$s(Kx) \leq 2|Kx| \leq \frac{1}{2}C_{58}|x|.$$

In virtue of these two inequalities, we deduce

$$e^{2s(Kx)} \leq e^{s(\lambda x)/2}, \qquad e^{2s(Kx)} \leq e^{C_{58}|x|} \tag{6.46}$$

for all $x \in \mathbb{R}^3$. Therefore, the choice of K enables us to relate the above exponential term with the exponential terms in the decay rates of both the steady-state and the purely periodic part of the vorticity fundamental solution; see Theorem 5.3.1 and Theorem 5.3.3.

Another aspect worth a comment is the second term in the definition of $N_S^\varepsilon(z)$, which captures the decay of the purely periodic part of $\operatorname{curl} z$ and contains the factor $|x|^{9/2-\varepsilon}$ instead of $|x|^{9/2}$. In the end, this difference ensures that $\mathcal{F}_S$ becomes a contraction in the underlying function space for large S. However, after having shown $N_S^\varepsilon(u|_{\mathbb{T}\times B^S}) < \infty$, we can finally omit ε by using representation formula (6.33) again.

We further introduce the function spaces $\mathcal{M}_S$ and $\mathcal{N}_S^\varepsilon$ associated to these (semi-)norms and set

$$\mathcal{M}_S := \left\{ z \in \mathrm{W}^{1,1}_{\mathrm{loc}}(\mathbb{T} \times B^S) \mid \mathrm{M}_S(z) < \infty \right\},$$
$$\mathcal{N}_S^\varepsilon := \left\{ z \in \mathcal{M}_S \mid \mathrm{N}_S^\varepsilon(z) < \infty \right\}.$$

Next let us show that these are Banach spaces.

Lemma 6.2.3. *Let $S > 0$ and $\varepsilon \in (0, 1/4)$. Let $\mathcal{M}_S$ and $\mathcal{N}_S^\varepsilon$ be equipped with the norms $\|\cdot\|_{\mathcal{M}_S}$ and $\|\cdot\|_{\mathcal{N}_S^\varepsilon}$ defined by*

$$\|z\|_{\mathcal{M}_S} := \mathrm{M}_S(z), \qquad \|z\|_{\mathcal{N}_S^\varepsilon} := \mathrm{M}_S(z) + \mathrm{N}_S^\varepsilon(z),$$

respectively. Then $\mathcal{M}_S$ and $\mathcal{N}_S^\varepsilon$ are Banach spaces.

Proof. Clearly, $\|\cdot\|_{\mathcal{M}_S}$ and $\|\cdot\|_{\mathcal{N}_S^\varepsilon}$ define norms on $\mathcal{M}_S$ and $\mathcal{N}_S^\varepsilon$, respectively. Let (z_j) be a Cauchy sequence in $\mathcal{M}_S$ or $\mathcal{N}_S^\varepsilon$. Then (z_j) is also a Cauchy sequence in $\mathrm{W}^{1,\infty}(\mathbb{T} \times B^S)$, and thus possesses a limit $z \in \mathrm{W}^{1,\infty}(\mathbb{T} \times B^S)$. For $j \in \mathbb{N}$ define

$$f_j(t, x) := |x|(1 + s(x))\mathcal{P}z_j(x) + |x|^3 \mathcal{P}_\perp z_j(t, x),$$
$$g_j(t, x) := \left[|x|(1 + s(x))\right]^{3/2} \nabla \mathcal{P}z_j(x) + |x|^4 \nabla \mathcal{P}_\perp z_j(t, x)$$

for $(t, x) \in \mathbb{T} \times B^S$.

If (z_j) is a Cauchy sequence in $\mathcal{M}_S$, then (f_j) and (g_j) are Cauchy sequences in $\mathrm{L}^\infty(\mathbb{T}\times B^S)$, which possess limits f and g, respectively. Because (f_j), (g_j), (z_j) and (∇z_j) converge pointwise almost everywhere, we see that

$$f(t, x) = |x|(1 + s(x))\mathcal{P}z(x) + |x|^3 \mathcal{P}_\perp z(t, x),$$
$$g(t, x) = \left[|x|(1 + s(x))\right]^{3/2} \nabla \mathcal{P}z(x) + |x|^4 \nabla \mathcal{P}_\perp z(t, x)$$

for almost all $(t, x) \in \mathbb{T} \times B^S$. Now $f, g \in \mathrm{L}^\infty(\mathbb{T}\times B^S)$ implies $z \in \mathcal{M}_S$. This shows completeness of $\mathcal{M}_S$.

If (z_j) is a Cauchy sequence in $\mathcal{N}_S^\varepsilon$, it is a Cauchy sequence in $\mathcal{M}_S$ as well, and we conclude $z \in \mathcal{M}_S$ as above. We further define

$$h_j(t,x) = |x|^{3/2}\, e^{\frac{s(Kx)}{1+S}}\, \mathrm{curl}\, \mathcal{P}z_j(x) + |x|^{9/2-\varepsilon}\, e^{\frac{s(Kx)}{1+S}}\, \mathrm{curl}\, \mathcal{P}_\perp z_j(t,x).$$

Then, (h_j) is a Cauchy sequence in $\mathrm{L}^\infty(\mathbb{T} \times \mathrm{B}^S)$ and possesses a limit $h \in \mathrm{L}^\infty(\mathbb{T} \times \mathrm{B}^S)$. Because (h_j) and (∇z_j) converge pointwise almost everywhere, we conclude

$$h(t,x) = |x|^{3/2}\, e^{\frac{s(Kx)}{1+S}}\, \mathrm{curl}\, \mathcal{P}z(x) + |x|^{9/2-\varepsilon}\, e^{\frac{s(Kx)}{1+S}}\, \mathrm{curl}\, \mathcal{P}_\perp z(t,x),$$

and $h \in \mathrm{L}^\infty(\mathbb{T} \times \mathrm{B}^S)$ implies $z \in \mathcal{N}_S^\varepsilon$. Therefore, $\mathcal{N}_S^\varepsilon$ is complete. $\qquad\square$

In the following two sections we give estimates of $\mathcal{F}_S(z)$ and $\mathcal{H}_S$ with respect to the above (semi-)norms. In the end, these estimates imply that $z \mapsto \mathcal{F}_S(z) + \mathcal{H}_S$ is a contractive self-mapping when we choose S sufficiently large.

6.2.4 Estimates of $\mathcal{H}_S$

Here, we collect estimates of

$$\mathcal{H}_S = \big(\Gamma^\lambda * \big[f - \chi_S\, \mathrm{curl}\, u \wedge u\big]\big)|_{\mathbb{T}\times\mathrm{B}^S} = \big(\Gamma^\lambda * \big[f + \chi_S\mathcal{A}(u)\big]\big)|_{\mathbb{T}\times\mathrm{B}^S}.$$

Before we derive the required estimates, we establish the following lemma. While its proof is elementary, it leads to the appearance of the term $1 + S$ in the definition of $\mathrm{N}_S^\varepsilon(z)$ above.

Lemma 6.2.4. *Let $a \in \mathbb{R}$, $b > 0$ and $S > 0$. If $y \in \mathbb{R}$ with $|y| \le 2S$, then*

$$e^{-s(a(x-y))} \le e^{4|a|}\, e^{-\frac{s(ax)}{1+S}}, \tag{6.47}$$

$$e^{-b|x-y|} \le e^{2b}\, e^{-\frac{b|x|}{1+S}}. \tag{6.48}$$

Proof. First note that for $|y| \le 2S$ we have

$$\frac{s(ay)}{1+S} \le \frac{2|a||y|}{1+S} \le \frac{4|a|S}{1+S} \le 4|a|.$$

Together with $s(a(x-y)) \ge s(ax) - s(ay)$, this implies

$$e^{-s(a(x-y))} \le e^{-\frac{s(a(x-y))}{1+S}} \le e^{-\frac{s(ax)}{1+S}}\, e^{\frac{s(ay)}{1+S}} \le e^{-\frac{s(ax)}{1+S}}\, e^{4|a|},$$

Similarly, we have $|y|/(1+S) \le 2S/(1+S) \le 2$, which for $b > 0$ implies

$$e^{-b|x-y|} \le e^{-\frac{b|x-y|}{1+S}} \le e^{-\frac{b|x|}{1+S}}\, e^{\frac{b|y|}{1+S}} \le e^{-\frac{b|x|}{1+S}}\, e^{2b}.$$

We have thus established (6.47) and (6.48). $\qquad\square$

In the following two lemmas, we show that the norm of $\mathcal{H}_S$ in both $\mathcal{M}_S$ and $\mathcal{N}_S^\varepsilon$ is bounded by a constant that is independent of $S \geq 2S_0$. Both proofs strongly rely on the convolution estimates from Theorem 5.1.7 and Theorem 5.2.8.

Lemma 6.2.5. *There exists a constant $C_{75} > 0$ such that for all $S \in [2S_0, \infty)$ we have*
$$\mathrm{M}_S(\mathcal{H}_S) \leq C_{75}.$$

Proof. Clearly, $f \in C_0^\infty(\mathbb{T} \times \mathbb{R}^3)$ implies $|f(t, x)| \leq c_0$ for all $(t, x) \in \mathbb{T} \times \mathbb{R}^3$. Combining this with the decay estimates of u and ∇u from Theorem 6.1.9, we thus obtain

$$\left| \mathcal{P}f(x) + \chi_S(x)\mathcal{A}_0(u)(x) \right| \leq c_1 \left[(1 + |x|)(1 + s(x)) \right]^{-5/2}, \qquad (6.49)$$
$$\left| \mathcal{P}_\perp f(t, x) + \chi_S(x)\mathcal{A}_\perp(u)(t, x) \right| \leq c_2 (1 + |x|)^{-9/2} \qquad (6.50)$$

for all $(t, x) \in \mathbb{T} \times \mathbb{R}^3$, where $\mathcal{A}_0$ and $\mathcal{A}_\perp$ are defined in (6.36) and (6.37). Exploiting the formulas (6.42) and (6.43) in combination with these estimates, Theorem 5.1.7 and Theorem 5.2.8 directly imply

$$\left| \mathcal{P}\mathcal{H}_S(x) \right| \leq c_3 \left[(1 + |x|)(1 + s(x)) \right]^{-1},$$
$$\left| \partial_j \mathcal{P}\mathcal{H}_S(x) \right| \leq c_4 \left[(1 + |x|)(1 + s(x)) \right]^{-3/2},$$
$$\left| \mathcal{P}_\perp \mathcal{H}_S(t, x) \right| \leq c_5 (1 + |x|)^{-3},$$
$$\left| \partial_j \mathcal{P}_\perp \mathcal{H}_S(t, x) \right| \leq c_6 (1 + |x|)^{-4}$$

for all $t \in \mathbb{T}$ and $|x| \geq S_0$. Collecting these, we arrive at the claimed estimate. $\qquad \square$

Lemma 6.2.6. *There exists a constant $C_{76} > 0$ such that for all $S \in [2S_0, \infty)$ we have*
$$\mathrm{N}_S^\varepsilon(\mathcal{H}_S) \leq C_{76}.$$

Proof. At first, consider $x \in \mathbb{R}^3$ with $|x| \geq 2S$. For $|y| \leq 7S/4$ we have

$$|x - y| \geq |x| - |y| \geq |x| - 7S/4 \geq |x| - 7|x|/8 = |x|/8 \geq S/4 \geq S_0/2.$$

From (5.61) and Lemma 6.2.4, we then conclude

$$\left| \nabla \phi_0^\lambda (x - y) \right| \leq c_0 \left(|x - y|^{-2} + |x - y|^{-3/2} s(\lambda(x - y))^{1/2} \right) e^{-s(\lambda(x-y))/2}$$
$$\leq c_1 \left(1 + |x - y|^{-3/2} \right) \left(1 + s(\lambda(x - y)) \right)^{-3/2} e^{-s(\lambda(x-y))/4}$$
$$\leq c_2 \left[(1 + |x - y|)(1 + s(\lambda(x - y))) \right]^{-3/2} e^{-\frac{s(\lambda x)}{4(1+S)}}.$$

In virtue of (6.49) and (6.44), we thus obtain

$$|\operatorname{curl}\mathcal{P}\mathcal{H}_S(x)| = \left| \int_{B_{7S/4}} \nabla\phi_0^\lambda(x-y) \wedge \left[\mathcal{P}f + \chi_S\mathcal{A}_0(u)\right](y)\,dy \right|$$

$$\leq c_3\, e^{-\frac{s(\lambda x)}{4(1+S)}} \int_{\mathbb{R}^3} \left[(1+|x-y|)(1+s(x-y))\right]^{-3/2}\left[(1+|y|)(1+s(y))\right]^{-5/2}\,dy$$

for $|x| \geq 2S \geq 4S_0$. Estimating the remaining integral with the help of Lemma A.2.2 and employing (6.46), we deduce

$$|\operatorname{curl}\mathcal{P}\mathcal{H}_S(x)| \leq c_4\, e^{-\frac{s(\lambda x)}{4(1+S)}}\,|x|^{-3/2} \leq c_5\, e^{-\frac{s(Kx)}{1+S}}\,|x|^{-3/2} \tag{6.51}$$

for $|x| \geq 2S$. If $x \in \mathbb{R}^3$ with $S \leq |x| \leq 2S$, then Lemma 6.2.5 yields

$$|\operatorname{curl}\mathcal{P}\mathcal{H}_S(x)| \leq c_6|\nabla\mathcal{P}\mathcal{H}_S(x)| \leq c_7\left[(1+|x|)(1+s(x))\right]^{-3/2} \leq c_8|x|^{-3/2}.$$

Since $|x| \leq 2S$ implies $s(Kx)/(1+S) \leq 2|Kx|/(1+S) \leq 4|K|S/(1+S) \leq 4|K|$, we have $1 \leq e^{4|K|}\,e^{-s(Kx)/(1+S)}$, so that (6.51) also holds for $S \leq |x| \leq 2S$.

Now let us turn to $\operatorname{curl}\mathcal{P}_\perp\mathcal{H}_S$. From (5.66) and (6.48) we conclude

$$\int_{\mathbb{T}} |\nabla\phi_\perp^\lambda(t-s, x-y)|\,ds \leq c_9|x-y|^{-5/2}\,e^{-C_{58}|x-y|}$$

$$\leq c_{10}|x-y|^{-5/2}\,e^{-\frac{C_{58}|x-y|}{2}}\,e^{-\frac{C_{58}|x|}{2(1+S)}},$$

so that (6.50) and (6.45) lead to

$$|\operatorname{curl}\mathcal{P}_\perp\mathcal{H}_S(t,x)| \leq c_{11} \int_{B_{7S/4}} \int_{\mathbb{T}} |\nabla\phi_\perp^\lambda(t-s, x-y)|\,\big|\mathcal{P}_\perp f + \chi_S\mathcal{A}_\perp(u)\big|(s,y)\,ds\,dy$$

$$\leq c_{12}\, e^{-\frac{C_{58}|x|}{2(1+S)}} \int_{\mathbb{R}^3} |x-y|^{-5/2}\,e^{-C_{58}|x-y|/2}(1+|y|)^{-9/2}\,dy.$$

The remaining integral can now be estimated with Lemma A.2.3. Further using (6.46), we end up with

$$|\operatorname{curl}\mathcal{P}_\perp\mathcal{H}_S(t,x)| \leq c_{13}\, e^{-\frac{C_{58}|x|}{2(1+S)}}\,|x|^{-9/2} \leq c_{14}\, e^{-\frac{s(Kx)}{1+S}}\,|x|^{-9/2+\varepsilon}$$

for $|x| \geq S \geq 2S_0$ and $t \in \mathbb{T}$. A combination of this estimate with (6.51) finishes the proof. $\qquad\square$

6.2.5 Estimates of $\mathcal{F}_S(z)$

In the following two lemmas, we give estimates of

$$\mathcal{F}_S(z) = \left(\Gamma^\lambda * \left[-(1 - \chi_S)\,\mathrm{curl}\,z \wedge z\right]\right)|_{\mathrm{T} \times \mathrm{B}^S} = \left(\Gamma^\lambda * \left[(1 - \chi_S)\mathcal{A}(z)\right]\right)|_{\mathrm{T} \times \mathrm{B}^S},$$

analogously to those previously established for $\mathcal{H}_S$. In contrast to $\mathcal{H}_S$, the term $\mathcal{F}(z)$ depends on the (unknown) function z, which is why estimates of differences for distinct arguments are also required for the fixed-point argument. Note that, in order to eventually obtain a contractive mapping for large S, we always factor out the term $S^{-\varepsilon}$.

Lemma 6.2.7. *There exists a constant $C_{77} > 0$ such that for all $S \in [2S_0, \infty)$ and all $z_1, z_2 \in \mathcal{M}_S$ we have*

$$M_S(\mathcal{F}_S(z_1)) \le C_{77} S^{-\varepsilon} M_S(z_1)^2, \tag{6.52}$$

$$M_S(\mathcal{F}_S(z_1) - \mathcal{F}_S(z_2)) \le C_{77} S^{-\varepsilon} \left(M_S(z_1) + M_S(z_2)\right) M_S(z_1 - z_2). \tag{6.53}$$

Proof. Let $z \in \mathcal{M}_S$ and recall the definition of $\mathcal{A}_0(z)$ and $\mathcal{A}_\perp(z)$ in (6.36) and (6.37). We immediately deduce

$$\left|(1 - \chi_S(x))\mathcal{A}_0(z)(x)\right| \le c_0 M_S(z)^2 (1 - \chi_S(x))\left[(1 + |x|)(1 + s(x))\right]^{-5/2}$$
$$\le c_1 S^{-\varepsilon} M_S(z)^2 (1 + |x|)^{-5/2+\varepsilon}(1 + s(x))^{-5/2},$$

$$\left|(1 - \chi_S(x))\mathcal{A}_\perp(z)(t, x)\right| \le c_2 M_S(z)^2 (1 - \chi_S(x))(1 + |x|)^{-9/2}$$
$$\le c_3 S^{-\varepsilon} M_S(z)^2 (1 + |x|)^{-9/2+\varepsilon}$$

for $|x| \ge S$. Recalling (6.38) and (6.39), from these estimates, Theorem 5.1.7 and Theorem 5.2.8 we conclude

$$\left|\mathcal{P}\mathcal{F}_S(z)(x)\right| \le c_4 S^{-\varepsilon} M_S(z)^2 \left[(1 + |x|)(1 + s(x))\right]^{-1},$$
$$\left|\partial_j \mathcal{P}\mathcal{F}_S(z)(x)\right| \le c_5 S^{-\varepsilon} M_S(z)^2 \left[(1 + |x|)(1 + s(x))\right]^{-3/2},$$
$$\left|\mathcal{P}_\perp \mathcal{F}_S(z)(t, x)\right| \le c_6 S^{-\varepsilon} M_S(z)^2 (1 + |x|)^{-3},$$
$$\left|\partial_j \mathcal{P}_\perp \mathcal{F}_S(z)(t, x)\right| \le c_7 S^{-\varepsilon} M_S(z)^2 (1 + |x|)^{-4}.$$

Collecting these estimates, we arrive at (6.52). Estimate (6.53) follows in the same fashion. $\qquad \square$

Lemma 6.2.8. *There exists a constant $C_{78} > 0$ such that for all $S \in [2S_0, \infty)$ and all $z_1, z_2 \in \mathcal{N}_S^\varepsilon$ we have*

$$N_S^\varepsilon(\mathcal{F}_S(z_1)) \le C_{78} S^{-\varepsilon} M_S(z_1) N_S^\varepsilon(z_1), \tag{6.54}$$

$$N_S^\varepsilon(\mathcal{F}_S(z_1) - \mathcal{F}_S(z_2)) \le C_{78} S^{-\varepsilon} \left(\|z_1\|_{\mathcal{N}_S^\varepsilon} + \|z_2\|_{\mathcal{N}_S^\varepsilon}\right) \|z_1 - z_2\|_{\mathcal{N}_S^\varepsilon}, \tag{6.55}$$

where $\|z\|_{\mathcal{N}_S^\varepsilon} = M_S(z) + N_S^\varepsilon(z)$.

Proof. Let $z \in \mathcal{N}_S^\varepsilon$. Recalling the definition of $\|\cdot\|_{\mathcal{N}_S^\varepsilon}$, we then have

$$|(1 - \chi_S(x))\mathcal{A}_0(z)(x)|$$
$$\leq c_0 M_S(z) N_S^\varepsilon(z)(1 - \chi_S(x))|x|^{-5/2}(1 + s(x))^{-1} e^{-\frac{s(Kx)}{1+S}} \qquad (6.56)$$
$$\leq c_1 S^{-\varepsilon} M_S(z) N_S^\varepsilon(z)|x|^{-5/2+\varepsilon}(1 + s(x))^{-1} e^{-\frac{s(Kx)}{1+S}},$$
$$|(1 - \chi_S(x))\mathcal{A}_\perp(z)(t, x)|$$
$$\leq c_2 M_S(z) N_S^\varepsilon(z)(1 - \chi_S(x))|x|^{-9/2} e^{-\frac{s(Kx)}{1+S}} \qquad (6.57)$$
$$\leq c_3 S^{-\varepsilon} M_S(z) N_S^\varepsilon(z)|x|^{-9/2+\varepsilon} e^{-\frac{s(Kx)}{1+S}}$$

for $|x| \geq S$. From (5.61) we conclude

$$|\nabla \phi_0^\lambda(x - y)| \leq c_4 \begin{cases} |x - y|^{-2} e^{-\frac{s(\lambda(x-y))}{4}} & \text{if } |x - y| \leq S_0, \\ [(1 + |x - y|)s(\lambda(x - y))]^{-3/2} e^{-\frac{s(\lambda(x-y))}{4}} & \text{if } |x - y| \geq S_0. \end{cases}$$

Therefore, exploiting the representation formula (6.40), we can employ (6.56) to estimate

$$|\text{curl}\, \mathcal{PF}_S(z)(x)| = \left| \int_{\mathbb{R}^3} \nabla \phi_0^\lambda(x - y) \wedge [(1 - \chi_S(y))\mathcal{A}_0(z)(y)]\, dy \right|$$
$$\leq c_5 S^{-\varepsilon} M_S(z) N_S^\varepsilon(z)(I_1 + I_2),$$

where

$$I_1 := \int_{B^S \cap B_{S_0}(x)} |x - y|^{-2} e^{-\frac{s(\lambda(x-y))}{4}} |y|^{-5/2+\varepsilon}(1 + s(y))^{-1} e^{-\frac{s(Ky)}{1+S}}\, dy,$$

$$I_2 := \int_{B^S \cap B^{S_0}(x)} [|x - y|s(\lambda(x - y))]^{-3/2} e^{-\frac{s(\lambda(x-y))}{4}}$$
$$\times |y|^{-5/2+\varepsilon}(1 + s(y))^{-1} e^{-\frac{s(Ky)}{1+S}}\, dy.$$

To give estimates of these integrals, we first note that by the elementary estimate $s(\lambda(x - y)) \geq s(\lambda x) - s(\lambda y)$ and (6.46), we have

$$e^{-\frac{s(\lambda(x-y))}{4}} e^{-\frac{s(Ky)}{1+S}} \leq e^{-\frac{s(\lambda x)}{4(1+S)}} e^{\frac{s(\lambda y)}{4(1+S)}} e^{-\frac{s(Ky)}{1+S}} \leq e^{-\frac{s(Kx)}{1+S}} \qquad (6.58)$$

for all $x, y \in \mathbb{R}^3$. On the one hand, exploiting this estimate and that $|x - y| \leq S_0 \leq |x|/2$ implies $|y| \geq |x| - |x - y| \geq |x| - S_0 \geq |x|/2$, we conclude

$$I_1 \leq c_6 e^{-\frac{s(Kx)}{1+S}} |x|^{-5/2+\varepsilon} \int_{B_{S_0}(x)} |x - y|^{-2}\, dy \leq c_7 e^{-\frac{s(Kx)}{1+S}} |x|^{-3/2}$$

for $|x| \geq S \geq 2S_0$. On the other hand, due to (6.58) and the fact that $|y| \geq S \geq 2S_0$ implies $|y| \geq c_8(1 + |y|)$, we obtain

$$I_2 \leq c_9 \, e^{-\frac{s(Kx)}{1+S}} \int_{\mathbb{R}^3} \left[(1 + |x - y|) s(x - y) \right]^{-3/2} (1 + |y|)^{-5/2+\varepsilon} (1 + s(y))^{-1} \, dy$$

$$\leq c_{10} \, e^{-\frac{s(Kx)}{1+S}} \, |x|^{-3/2}$$

by Lemma A.2.2. From the estimates of I_1 and I_2 we deduce

$$|\operatorname{curl} \mathcal{P} \mathcal{F}_S(z)(x)| \leq c_{11} S^{-\varepsilon} M_S(z) N_S^\varepsilon(z) \, e^{-\frac{s(Kx)}{1+S}} \, |x|^{-3/2}.$$

Now let us turn to the purely periodic part $\mathcal{P}_\perp \mathcal{F}_S(z)$. From (5.66) (with $q = 1$ and $\gamma = 1/4$) we conclude

$$\int_{\mathbb{T}} |\nabla \phi_\perp^\lambda(t - s, x - y)| \, ds \leq c_{12} |x - y|^{-5/2} \, e^{-C_{58}|x-y|}.$$

With formula (6.41) and estimate (6.57) we thus obtain

$$|\operatorname{curl} \mathcal{P}_\perp \mathcal{F}_S(z)(t, x)|$$

$$= \left| \int_{\mathbb{T}} \int_{\mathbb{R}^3} \nabla \phi_\perp^\lambda(t - s, x - y) \wedge \left[(1 - \chi_S(y)) \mathcal{A}_\perp(z)(y) \right] dy \, ds \right|$$

$$\leq c_{13} S^{-\varepsilon} M_S(z) N_S^\varepsilon(z) \int_{B^S} |x - y|^{-5/2} \, e^{-C_{58}|x-y|} \, |y|^{-9/2+\varepsilon} \, e^{-\frac{s(Ky)}{1+S}} \, dy.$$

By (6.46) we have

$$e^{-\frac{C_{58}|x-y|}{2}} e^{-\frac{s(Ky)}{1+S}} \leq e^{-s(K(x-y))} e^{-\frac{s(Ky)}{1+S}} \leq e^{-\frac{s(K(x-y))}{1+S}} e^{-\frac{s(Ky)}{1+S}} \leq e^{-\frac{s(Kx)}{1+S}}.$$

This yields

$$|\operatorname{curl} \mathcal{P}_\perp \mathcal{F}_S(z)(t, x)|$$

$$\leq c_{14} S^{-\varepsilon} M_S(z) N_S^\varepsilon(z) \, e^{-\frac{s(Kx)}{1+S}} \int_{\mathbb{R}^3} |x - y|^{-5/2} \, e^{-\frac{C_{58}|x-y|}{2}} (1 + |y|)^{-9/2+\varepsilon} \, dy.$$

Employing Lemma A.2.3 to estimate the remaining integral, we end up with

$$|\operatorname{curl} \mathcal{P}_\perp \mathcal{F}_S(z)(t, x)| \leq c_{15} S^{-\varepsilon} M_S(z) N_S^\varepsilon(z) \, e^{-\frac{Ks(Kx)}{1+S}} \, |x|^{-9/2+\varepsilon}.$$

In total, we have thus shown (6.54). Estimate (6.55) is derived in the same way. $\qquad \square$

6.2.6 Spatial Decay Estimates

After the preparatory results from the previous subsection, we now prove the existence of a function $z \in \mathcal{N}_S^\varepsilon$ satisfying the fixed-point equation

$$z = \mathcal{F}_S(z) + \mathcal{H}_S$$

provided $S \geq 2S_0$ is chosen sufficiently large. Afterwards, we show uniqueness of this fixed point in the function class $\mathcal{M}_S$. Since $u|_{\mathbb{T} \times B^S}$ is another solution to this fixed-point equation and belongs to $\mathcal{M}_S$ by Theorem 6.1.9, we then conclude that z coincides with $u|_{\mathbb{T} \times B^S}$. This yields the desired decay rate of the vorticity field up to a factor $|x|^{-\varepsilon}$ for the purely periodic part. Returning to the representation formula (6.33), we can finally omit this factor.

To begin with, for $S \in [2S_0, \infty)$ consider the closed subset

$$\mathcal{B}_S := \left\{ z \in \mathcal{N}_S^\varepsilon \mid \|z\|_{\mathcal{N}_S^\varepsilon} \leq C_{75} + C_{76} + 1 \right\}$$

of the Banach space $\mathcal{N}_S^\varepsilon$. Choose $S_1 \in [2S_0, \infty)$ so large that for all $S \in [S_1, \infty)$ we have

$$(C_{77} + C_{78})(C_{75} + C_{76} + 1)^2 S^{-\varepsilon} \leq 1,$$

$$(C_{77} + C_{78})(C_{75} + C_{76} + 1) S^{-\varepsilon} \leq \frac{1}{4}.$$

Then we obtain the following.

Corollary 6.2.9. *For any $S \in [S_1, \infty)$ there is a function $z_S \in \mathcal{B}_S$ with $z_S = \mathcal{F}_S(z_S) + \mathcal{H}_S$.*

Proof. We define the mapping

$$F_S : \mathcal{B}_S \to \mathcal{B}_S, \qquad F_S(z) := \mathcal{F}_S(z) + \mathcal{H}_S.$$

By the Lemma 6.2.5, Lemma 6.2.6, Lemma 6.2.7 and Lemma 6.2.8 and the choice of S_1, this is a well-defined contractive self-mapping for any $S \geq S_1$. The contraction mapping principle thus implies the existence of a fixed point $z_S \in \mathcal{B}_S$ of F_S, that is, of a function z_S with the asserted properties. □

Next we show that this function z_S coincides with $u|_{\mathbb{T} \times B^S}$ for S sufficiently large in order to obtain the following intermediate result.

Lemma 6.2.10. *There exists $S_2 \in [S_1, \infty)$ such that for all $S \in [S_2, \infty)$ we have*

$$|\operatorname{curl} \mathcal{P}u(x)| \le (C_{75} + C_{76} + 1)|x|^{-3/2} e^{-\frac{s(Kx)}{1+S}},$$

$$|\operatorname{curl} \mathcal{P}_\perp u(t, x)| \le (C_{75} + C_{76} + 1)|x|^{-9/2+\varepsilon} e^{-\frac{s(Kx)}{1+S}}$$

for all $t \in \mathbb{T}$ and $x \in B^S$.

Proof. For $S \ge 2S_0$ we set $U_S := u|_{\mathbb{T}\times B^S}$. By Theorem 6.1.9 we know $U_S \in \mathcal{M}_S$ with $\mathrm{M}_S(U) \le C_{64}$, and by (6.35) we have $U_S = \mathcal{F}_S(U_S) + \mathcal{H}_S$ for any $S \ge 2S_0$. Now let $S \ge S_1$ and let $z_S \in \mathcal{B}_S$ be the function from Corollary 6.2.9. Then Lemma 6.2.7 implies

$$\begin{aligned} \mathrm{M}_S(z_S - U_S) &= \mathrm{M}_S(\mathcal{F}_S(z_S) - \mathcal{F}_S(U_S)) \\ &\le C_{77} S^{-\varepsilon} \big(\mathrm{M}_S(z_S) + \mathrm{M}_S(U_S) \big) \mathrm{M}_S(z_S - U_S) \\ &\le C_{77} S^{-\varepsilon} \big(C_{75} + C_{76} + 1 + C_{64} \big) \mathrm{M}_S(z_S - U_S). \end{aligned}$$

Choosing $S_2 \in [S_1, \infty)$ such that for all $S \in [S_2, \infty)$ we have

$$C_{77} S^{-\varepsilon} \big(C_{75} + C_{76} + 1 + C_{64} \big) \le \frac{1}{2},$$

we conclude $\mathrm{M}_S(z_S - U_S) \le \mathrm{M}_S(z_S - U_S)/2$ and hence $\mathrm{M}_S(z_S - U_S) = 0$ for all $S \in [S_2, \infty)$. This implies $z_S = U_S = u|_{\mathbb{T}\times B^S}$. In particular, we have $\mathrm{N}_S^\varepsilon(u|_{\mathbb{T}\times B^S}) = \mathrm{N}_S^\varepsilon(z_S) \le C_{75} + C_{76} + 1$ for all $S \in [S_2, \infty)$. This completes the proof. $\qquad\square$

Another application of the convolution formula (6.33) enables us to omit the term ε in the estimate of $\operatorname{curl} \mathcal{P}_\perp u$. We can thus prove the main theorem of this chapter.

Theorem 6.2.11. *Let $\lambda \ne 0$ and $f \in C_0^\infty(\mathbb{T}\times\mathbb{R}^3)^3$, and let u be a weak time-periodic solution to (6.2) in the sense of Definition 6.1.1, which satisfies (6.3). Then there exist constants C_{79} and $\alpha = \alpha(\lambda, \mathcal{T}) > 0$ such that the estimates*

$$|\operatorname{curl} \mathcal{P}u(x)| \le C_{79}(1 + |x|)^{-3/2} e^{-\alpha s(\lambda x)}, \tag{6.59}$$

$$|\operatorname{curl} \mathcal{P}_\perp u(t, x)| \le C_{79}(1 + |x|)^{-9/2} e^{-\alpha s(\lambda x)} \tag{6.60}$$

hold for all $t \in \mathbb{T}$ and $x \in \mathbb{R}^3$.

Proof. We decompose $u = v + w$ into steady-state part $v := \mathcal{P}u$ and purely periodic part $w := \mathcal{P}_\perp u$. By Lemma 6.1.7 we have $\operatorname{curl} u \in C^\infty(\mathbb{T} \times \mathbb{R}^3)$. Therefore, $\operatorname{curl} u$ is bounded on $\mathbb{T} \times B_{S_2}$ with S_2 from Lemma 6.2.10. Since $s(Kx)/(1+S_2) \le 2|Kx|/(1+S_2) \le 2|K|$ for $|x| \le S_2$, we conclude

$$|\operatorname{curl} v(x)| \le c_0(1+|x|)^{-3/2} e^{-\frac{s(Kx)}{1+S_2}},$$

$$|\operatorname{curl} w(t,x)| \le c_1(1+|x|)^{-9/2+\varepsilon} e^{-\frac{s(Kx)}{1+S_2}}$$

for $|x| \le S_2$. Combining these estimates with those from Lemma 6.2.10 (with $S = S_2$), we deduce

$$
\begin{aligned}
|\operatorname{curl} v(x)| &\le c_2(1+|x|)^{-3/2} e^{-\alpha s(\lambda x)}, \\
|\operatorname{curl} w(t,x)| &\le c_3(1+|x|)^{-9/2+\varepsilon} e^{-\alpha s(\lambda x)}
\end{aligned}
\tag{6.61}
$$

for all $(t,x) \in \mathbb{T} \times \mathbb{R}^3$, where $\alpha = (\lambda(1+S_2))^{-1} K$. In particular, this implies (6.59), and for (6.60) it remains to remove ε in the second inequality. Due to $f \in C_0^\infty(\mathbb{T} \times \mathbb{R}^3)$ and Theorem 6.1.9, the estimates in (6.61) further yield

$$\left| \mathcal{P}_\perp f(s,y) - \operatorname{curl} v(y) \wedge w(s,y) - \operatorname{curl} w(s,y) \wedge v(y) - \mathcal{P}_\perp[\operatorname{curl} w \wedge w](s,y) \right|$$
$$\le c_4(1+|y|)^{-9/2} e^{-\alpha s(\lambda y)}$$

for all $(t,x) \in \mathbb{T} \times \mathbb{R}^3$. Moreover, by Theorem 5.3.3 we have

$$\int_\mathbb{T} |\nabla \phi_\perp^\lambda(t-s,x-y)| \, ds \le c_5 |x-y|^{-5/2} e^{-C_{58}|x-y|}.$$

Using these estimates in the representation formula (6.33), we conclude

$$|\operatorname{curl} w(t,x)| \le c_6 \int_{\mathbb{R}^3} |x-y|^{-5/2} e^{-C_{58}|x-y|} (1+|y|)^{-9/2} e^{-\alpha s(\lambda y)} \, dy.$$

Due to $2s(Kx) \le C_{58}|x|$, we have

$$\frac{1}{2} C_{58}|x-y| + \alpha s(\lambda y) \ge s(K(x-y)) + \frac{s(Ky)}{1+S_2} \ge \frac{s(Kx)}{1+S_2} = \alpha s(\lambda x),$$

and we can obtain

$$|\operatorname{curl} w(t,x)| \le c_7 e^{-\alpha s(\lambda x)} \int_{\mathbb{R}^3} |x-y|^{-5/2} e^{-C_{58}|x-y|/2} (1+|y|)^{-9/2} \, dy.$$

We estimate the remaining integral with Lemma A.2.3, which leads to

$$|\operatorname{curl} w(t,x)| \le c_8 e^{-\alpha s(\lambda x)} |x|^{-9/2}.$$

Since $w \in C^\infty(\mathbb{T} \times \mathbb{R}^3)$, this shows (6.60) and completes the proof. $\qquad\square$

6.3 Spatial Decay in an Exterior Domain

Here we consider some applications of the previous results to the case of a Navier–Stokes flow in an exterior domain in two different configurations. First, we consider a time-periodic flow past a body moving inside a fixed region. By means of a cut-off procedure, we show that the corresponding velocity and vorticity fields show the same decay properties as we established in Theorem 6.1.9 and Theorem 6.2.11 for a flow in the whole space. Secondly, we consider the flow past a rotating body, which can be seen as a special case of the previous problem, but we consider steady-state solutions in a frame attached to the body. By means of a suitable coordinate transform, we reduce this problem to the previous one and derive asymptotic properties.

6.3.1 Time-Periodic Flow Past a Moving Body

Here we consider the viscous flow past a body $\mathcal{B}$ that moves inside the three-dimensional whole space. Let $\Omega(t)$ denote the region occupied by the fluid at time $t \in \mathbb{R}$. The flow past $\mathcal{B}$ is then governed by the equations

$$\begin{cases} \rho(\partial_t v + v \cdot \nabla v) = \mu \Delta v - \nabla p & \text{in } \bigcup_{t \in \mathbb{R}} \{t\} \times \Omega(t), \\ \operatorname{div} v = 0 & \text{in } \bigcup_{t \in \mathbb{R}} \{t\} \times \Omega(t), \\ \lim_{|x| \to \infty} v(t, x) = v_\infty & \text{for } t \in \mathbb{R}. \end{cases} \tag{6.62}$$

Here v and p denote velocity field and pressure field, and v_∞ denotes a constant inflow velocity "at infinity". Moreover, μ and ρ denote constant viscosity and density of the fluid. In the following, we only consider the case $v_\infty \neq 0$ and, by a simple change of coordinates, we may assume that v_∞ is directed along the negative x_1 axis, that is, $v_\infty = -|v_\infty| e_1$. Moreover, we assume that the motion of the body $\mathcal{B}$ takes place in a bounded region, say, inside the ball B_{R_1} for a fixed radius $R_1 > 0$, and that the fluid flow exterior of this region is time periodic, that is,

$$v(t + \mathcal{T}, x) = v(t, x), \quad p(t + \mathcal{T}, x) = p(t, x) \qquad \text{for } (t, x) \in \mathbb{R} \times B^{R_1}.$$

For example, this situation occurs when the body $\mathcal{B}$ oscillates or rotates with fixed angular velocity. Observe that we did not specify any boundary conditions of v at the boundary $\partial \Omega(t)$, and to ensure existence of a time-periodic solution (v, p) these would have to be suitably chosen. However, these boundary values are not relevant for the result presented here, and we restrict our consideration to the functions on the set $\mathbb{R} \times B^{R_1}$ in the following. Moreover, since v and p are time periodic on this domain, we can transform the equations into a torus setting.

We introduce non-dimensional coordinates. Let $|v_\infty|$ serve as a characteristic velocity, and let the diameter d of the body $\mathcal{B}$ serve as a characteristic length. We define the Reynolds number $\lambda := \rho d |v_\infty|/\mu$, and the non-dimensional space and time variables $x' = x/d$ and $t' = \mu t/(\rho d^2)$. The corresponding radius and time period are $R_0 = R_1/d$ and $\mathcal{T}' = \mu T/(\rho d^2)$. Recall the quotient mapping $\pi \colon \mathbb{R} \to \mathbb{T}$ for $\mathbb{T} = \mathbb{R}/\mathcal{T}'\mathbb{Z}$ and the corresponding representation mapping $\Pi \colon \mathbb{T} \to [0, \mathcal{T}')$. We introduce the dimensionless velocity u and pressure $\mathfrak{p}$ by

$$u(t', x') = \frac{\rho d}{\mu} \big(v(\Pi(t), x) - v_\infty \big), \qquad \mathfrak{p}(t', x') = \frac{\rho d^2}{\mu} p(\Pi(t), x).$$

for $(t', x') \in \mathbb{T} \times B^{R_0}$. Omitting the primes, from (6.62) we then obtain

$$\begin{cases} \partial_t u - \lambda \partial_1 u + u \cdot \nabla u = \Delta u - \nabla \mathfrak{p} & \text{in } \mathbb{T} \times B^{R_0}, \\ \operatorname{div} u = 0 & \text{in } \mathbb{T} \times B^{R_0}, \\ \lim_{|x| \to \infty} u(t, x) = 0 & \text{for } t \in \mathbb{T}. \end{cases} \qquad (6.63)$$

Clearly, spatially asymptotic properties of u are equally valid for v. We thus restrict our investigation to (6.63), for which we obtain the following theorem.

Theorem 6.3.1. *Let $\lambda \neq 0$ and let $(u, \mathfrak{p})$ be a solution to (6.63), and assume that there exists $R > R_0$ such that $(u, \mathfrak{p}) \in C^\infty(\mathbb{T} \times \overline{B^R})^{3+1}$ and*

$$u \in L^2(\mathbb{T}; D^{1,2}(B^R))^3, \quad \mathcal{P}u \in L^6(B^R)^3, \quad \mathcal{P}_\perp u \in L^\infty(\mathbb{T}; L^2(B^R))^3.$$

Then there exists a constant $C_{80} > 0$ and $\alpha > 0$ such that

$$|\mathcal{P}u(x)| \leq C_{80}\big[|x|\big(1 + s(\lambda x)\big)\big]^{-1}, \qquad |\mathcal{P}_\perp u(t, x)| \leq C_{80}|x|^{-3},$$

$$|\nabla \mathcal{P}u(x)| \leq C_{80}\big[|x|\big(1 + s(\lambda x)\big)\big]^{-\frac{3}{2}}, \qquad |\nabla \mathcal{P}_\perp u(t, x)| \leq C_{80}|x|^{-4},$$

$$|\operatorname{curl} \mathcal{P}u(x)| \leq C_{80}|x|^{-3/2}\, e^{-\alpha s(\lambda x)}, \qquad |\operatorname{curl} \mathcal{P}_\perp u(t, x)| \leq C_{80}|x|^{-9/2}\, e^{-\alpha s(\lambda x)}$$

for all $t \in \mathbb{T}$ and $x \in B^R$.

Proof. Clearly, the asserted estimates hold for $t \in \mathbb{T}$ and $R \le |x| \le 3R$ since u is smooth. Let $\chi \in C_0^\infty(\mathbb{R}^3)$ be a cut-off function with $\chi(x) = 1$ for $|x| \le 2R$ and $\chi(x) = 0$ for $|x| \ge 3R$. We set

$$w := (1 - \chi)u + \mathfrak{B}(u \cdot \nabla \chi), \qquad \mathfrak{q} := (1 - \chi)\mathfrak{p},$$

where $\mathfrak{B}$ denotes the Bogovskiĭ operator from Theorem 2.4.2. From the regularity assumptions on $(u, \mathfrak{p})$ we conclude $(w, \mathfrak{q}) \in C^\infty(\mathbb{T} \times \mathbb{R}^3)^{3+1}$ and

$$w \in L^2(\mathbb{T}; D^{1,2}(\mathbb{R}^3))^3, \quad \mathcal{P}w \in L^6(\mathbb{R}^3)^3, \quad \mathcal{P}_\perp w \in L^\infty(\mathbb{T}; L^2(\mathbb{R}^3))^3.$$

Moreover, $(w, \mathfrak{q})$ satisfies

$$\begin{cases} \partial_t w - \lambda \partial_1 w + w \cdot \nabla w = \Delta w - \nabla \mathfrak{q} + f & \text{in } \mathbb{T} \times \mathbb{R}^3, \\ \operatorname{div} w = 0 & \text{in } \mathbb{T} \times \mathbb{R}^3, \\ \lim_{|x| \to \infty} w(t, x) = 0 & \text{for } t \in \mathbb{T} \end{cases}$$

for a function $f \in C_0^\infty(\mathbb{T} \times \mathbb{R}^3)$. In view of Remark 6.1.4, w satisfies the assumptions of Theorem 6.1.9 and Theorem 6.2.11 and is therefore subject to the pointwise estimates given there. Since $u = w$ for $|x| \ge 3R$, this shows the asserted estimates for u and completes the proof. $\qquad \square$

6.3.2 Steady-State Flow Around a Rotating Body

As a consequence of Theorem 6.3.1 we can derive decay properties of the solutions to the equations

$$\begin{cases} \omega(e_1 \wedge u - e_1 \wedge x \cdot \nabla u) - \lambda \partial_1 u + u \cdot \nabla u = \Delta u - \nabla \mathfrak{p} & \text{in } \Omega, \\ \operatorname{div} u = 0 & \text{in } \Omega, \\ \lim_{|x| \to \infty} u(x) = 0, \end{cases} \tag{6.64}$$

which describe the steady flow of a Navier–Stokes fluid around a body that translates with constant velocity λe_1, $\lambda > 0$, and rotates about the translation axis with angular velocity $\omega > 0$. Here Ω is the exterior domain occupied by the fluid flow. The functions $u \colon \Omega \to \mathbb{R}^3$ and $\mathfrak{p} \colon \Omega \to \mathbb{R}$ describe corresponding steady-state velocity and pressure fields. We conclude the following theorem that establishes pointwise estimates of weak solutions to (6.64).

Theorem 6.3.2. *Let $(u, \mathfrak{p}) \in (L^6(\Omega)^3 \cap D^{1,2}(\Omega)^3) \times L^2_{loc}(\Omega)$ be a distributional solution to (6.64), that is, $(u, \mathfrak{p})$ satisfies $\operatorname{div} u = 0$ and*

$$0 = \int_\Omega \left[\nabla u : \nabla \psi + \left(\omega(e_1 \wedge u - e_1 \wedge x \cdot \nabla u) - \lambda \partial_1 u + u \cdot \nabla u \right) \cdot \psi - \mathfrak{p} \operatorname{div} \psi \right] dx$$

for all $\psi \in C_0^\infty(\Omega)^3$. Further assume $e_1 \wedge u - e_1 \wedge x \cdot \nabla u \in L^2(\Omega)^3$. For every $R > \delta(\Omega^c)$ there exists a constant $C_{81} > 0$ such that

$$|u(x)| \leq C_{81} \left[|x| (1 + s(\lambda x)) \right]^{-1},$$

$$|\nabla u(x)| \leq C_{81} \left[|x| (1 + s(\lambda x)) \right]^{-\frac{3}{2}},$$

$$|\operatorname{curl} u(x)| \leq C_{81} |x|^{-3/2} e^{-\alpha s(\lambda x)}$$

for all $x \in B^R$.

Proof. Let $R > R_0 > \delta(\Omega^c)$ and let

$$Q_\omega(t) := \begin{pmatrix} 1 & 0 & 0 \\ 0 & \cos(\omega t) & -\sin(\omega t) \\ 0 & \sin(\omega t) & \cos(\omega t) \end{pmatrix}$$

be the matrix corresponding to the rotation with angular velocity ωe_1. Define the new variable $y = Q_\omega(t)x$ and set

$$U(t, y) := Q_\omega(t)u(Q_\omega(t)^\top y), \quad \mathfrak{P}(t, y) := \mathfrak{p}(Q_\omega(t)^\top y)$$

for all $t \in \mathbb{R}$ and $|y| \geq R_0$. Then U and $\mathfrak{P}$ are time periodic with period $\mathcal{T} = 2\pi/\omega$, and we can identify them with functions on $\mathbb{T} \times B^{R_0}$, where $\mathbb{T} = \mathbb{R}/\mathcal{T}\mathbb{Z}$ denotes the corresponding torus group. By the regularity result [42, Theorem XI.1.2], we have $(u, \mathfrak{p}) \in C^\infty(\Omega)^{3+1}$, which implies $(U, \mathfrak{P}) \in C^\infty(\mathbb{T} \times B^{R_0})^{3+1}$. Due to the identity

$$\partial_t U(t, y) = \omega Q_\omega(t) \left(e_1 \wedge u(x) - e_1 \wedge x \cdot \nabla u(x) \right), \quad (6.65)$$

with $y = Q_\omega(t)^\top x$, the pair $(U, \mathfrak{P})$ satisfies

$$\begin{cases} \partial_t U - \lambda \partial_1 U + U \cdot \nabla U = \Delta U - \nabla \mathfrak{P} & \text{in } \mathbb{T} \times B^{R_0}, \\ \operatorname{div} U = 0 & \text{in } \mathbb{T} \times B^{R_0}. \end{cases}$$

Moreover, U and $\mathfrak{P}$ are smooth and

$$\int_\mathbb{T} \int_{B^R} |\nabla U(t, x)|^2 \, dx dt = \int_\mathbb{T} \int_{B^R} |\nabla u(Q_\omega(t)^\top x)|^2 dx dt \leq |u|^2_{1,2},$$

$$\int_{B^R} |\mathcal{P}U(t, x)|^6 \, dx dt \leq \int_\mathbb{T} \int_{B^R} |u(Q_\omega(t)^\top x)|^6 \, dx dt \leq \|u\|^6_6.$$

This implies $U \in L^2(\mathbb{T}; D^{1,2}(B^R))^3$ and $\mathcal{P}U \in L^6(B^R)^3$. By (6.65), the assumption $e_1 \wedge u - e_1 \wedge x \cdot \nabla u \in L^2(\Omega)^3$ implies $\partial_t U \in L^2(\mathbb{T}; L^2(\Omega))^3$ and thus $\mathcal{P}_\perp U \in L^\infty(\mathbb{T}; L^2(\Omega))^3$. By Theorem 6.3.1 we thus conclude

$$|U(t,x)| \le c_0 \big[|x|(1 + s(\lambda x))\big]^{-1},$$

$$|\nabla U(t,x)| \le c_1 \big[|x|(1 + s(\lambda x))\big]^{-\frac{3}{2}},$$

$$|\operatorname{curl} U(t,x)| \le c_2 |x|^{-3/2} e^{-\alpha s(\lambda x)}$$

for all $(t,x) \in \mathbb{T} \times B^R$. The asserted estimates for u follow by a change of coordinates back to the frame attached to the body. □

Appendix

A.1 Estimates of Specific Functions

A.1.1 Hankel Functions

For $\nu \in \mathbb{C}$ the Hankel functions $H_\nu^{(1)}$ and $H_\nu^{(2)}$ of first and second kind are two particular linearly independent solutions to the Bessel differential equation

$$x^2 \frac{\mathrm{d}^2}{\mathrm{d}x^2} H_\nu^{(j)}(x) + x \frac{\mathrm{d}}{\mathrm{d}x} H_\nu^{(j)}(x) + (x^2 - \nu^2) H_\nu^{(j)}(x) = 0.$$

They are given by $H_\nu^{(1)} = J_\nu + iY_\nu$ and $H_\nu^{(2)} = J_\nu - iY_\nu$, where J_ν and Y_ν are the Bessel functions of first and second kind, given by

$$J_\nu(x) = \sum_{m=0}^{\infty} \frac{(-1)^m}{m! \, \Gamma(m + \nu + 1)} \left(\frac{x}{2}\right)^{2m+\nu}$$

169

and

$$Y_\nu(x) = \lim_{\mu \to \nu} \frac{J_\mu(x)\cos(\mu\pi) - J_{-\mu}(x)}{\sin(\mu\pi)}.$$

Note that the limit in the last identity can be omitted if ν is not an integer. Moreover, the modified Bessel functions I_ν and K_ν are defined by

$$I_\nu(x) = \sum_{m=0}^{\infty} \frac{1}{m!\,\Gamma(m+\nu+1)} \left(\frac{x}{2}\right)^{2m+\nu},$$

and

$$K_\nu(x) = \lim_{\mu \to \nu} \frac{\pi}{2} \frac{I_{-\mu}(x) - I_\mu(x)}{\sin(\mu\pi)}.$$

In the following, we focus in the study of the Hankel functions. First, we collect some of their well known properties.

Lemma A.1.1. *Hankel functions* $H_\nu^{(j)}$, $j = 1, 2$, *are analytic in* $\mathbb{C} \setminus \{0\}$ *with*

$$\forall \nu \in \mathbb{C}\ \forall z \in \mathbb{C} \setminus \{0\}: \quad \frac{d}{dz} H_\nu^{(j)}(z) = H_{\nu-1}^{(j)}(z) - \frac{\nu}{z} H_\nu^{(j)}(z), \quad (A.66)$$

and they satisfy the following estimates:

$$\forall \nu \in \mathbb{C}\ \forall \varepsilon > 0\ \exists C_{82} > 0\ \forall |z| \geq \varepsilon: \quad \left|H_\nu^{(j)}(z)\right| \leq C_{82} |z|^{-\frac{1}{2}} e^{(-1)^j \operatorname{Im} z}, \quad (A.67)$$

$$\forall \nu \in \mathbb{C}_+\ \forall R > 0\ \exists C_{83} > 0\ \forall |z| \leq R: \quad \left|H_\nu^{(j)}(z)\right| \leq C_{83} |z|^{-\nu}, \quad (A.68)$$

$$\forall 0 \leq R < 1\ \exists C_{84} > 0\ \forall |z| \leq R: \quad \left|H_0^{(j)}(z)\right| \leq C_{84} \left|\log(|z|)\right|. \quad (A.69)$$

Proof. The recurrence relation (A.66) is a well-know property of various Bessel functions; see for example [1, 9.1.27]. We refer to [1, 9.2.3] for the asymptotic behavior (A.67) of $H_\nu^{(j)}(z)$ as $|z| \to \infty$, and to [1, 9.1.9 and 9.1.8] for the asymptotic behavior (A.68) and (A.69) of $H_\nu^{(j)}(z)$ as $|z| \to 0$. $\qquad \square$

Next we study the function $x \mapsto H_\nu^{(j)}(a|x|)$ for $x \in \mathbb{R}^n$, $n \geq 2$ and a parameter $a \in \mathbb{C}$. We first derive a general formula for its derivative, which is based on the recurrence relation (A.66).

Lemma A.1.2. *Let* $\nu, a \in \mathbb{C}$, $n \in \mathbb{N}$ *and* $\alpha \in \mathbb{N}_0^n$. *Then*

$$D^\alpha\left[H_\nu^{(j)}(a|x|)\right] = \frac{1}{|x|^{2|\alpha|}} \sum_{\ell=0}^{|\alpha|} p_{\alpha,\ell}(x)\left(a|x|\right)^\ell H_{\nu-\ell}^{(j)}(a|x|) \quad (A.70)$$

for all $x \in \mathbb{R}^n \setminus \{0\}$, *where* $p_{\alpha,\ell} : \mathbb{R}^n \to \mathbb{R}$ *are polynomials in* x, *independent of* a, *such that* $\deg p_{\alpha,\ell} \leq |\alpha|$.

Proof. We show the statement inductively. For $\alpha = 0$, there is nothing to do. So assume that representation (A.70) holds for some $\alpha \in \mathbb{N}_0^n$. Then we compute

$$\partial_m \mathrm{D}^\alpha \big[H_\nu^{(j)}(a|x|) \big]$$

$$= \sum_{\ell=0}^{|\alpha|} \Bigg[\frac{-2|\alpha| x_m p_{\alpha,\ell}(x)}{|x|^{2|\alpha|+2}} (a|x|)^\ell H_{\nu-\ell}^{(j)}(a|x|) + \frac{\partial_m p_{\alpha,\ell}(x)}{|x|^{2|\alpha|}} (a|x|)^\ell H_{\nu-\ell}^{(j)}(a|x|)$$

$$+ \frac{p_{\alpha,\ell}(x)}{|x|^{2|\alpha|}} \ell (a|x|)^\ell \frac{x_m}{|x|^2} H_{\nu-\ell}^{(j)}(a|x|) + \frac{p_{\alpha,\ell}(x)}{|x|^{2|\alpha|}} (a|x|)^\ell \big(\frac{\mathrm{d}}{\mathrm{d}z} H_{\nu-\ell}^{(j)} \big)(a|x|) \frac{a x_m}{|x|} \Bigg]$$

$$= \sum_{\ell=0}^{|\alpha|} \Bigg[\big[+ \partial_m p_{\alpha,\ell}(x)|x|^2 + (\ell - \nu - 2|\alpha|) p_{\alpha,\ell}(x) x_m \big] (a|x|)^\ell H_{\nu-\ell}^{(j)}(a|x|)$$

$$+ p_{\alpha,\ell}(x) x_m (a|x|)^{\ell+1} H_{\nu-\ell-1}^{(j)}(a|x|) \Bigg] \frac{1}{|x|^{2|\alpha|+2}},$$

where we used the recurrence relation (A.66). Therefore, the function $\partial_m \mathrm{D}^\alpha \big[H_\nu^{(j)}(a|x|) \big]$ is of the claimed form, and the assertion follows by induction. $\qquad\square$

A combination of Lemma A.1.2 with the asymptotic behavior from (A.67) leads to the following decay estimate.

Lemma A.1.3. *Let $\nu \in \mathbb{C}$, $n \in \mathbb{N}$, $k \in \mathbb{N}_0$ and $\delta, \varepsilon > 0$. Then there exists a constant $C_{85} = C_{85}(n, k, \nu, \delta, \varepsilon) > 0$ such that*

$$\big| \nabla^k \big[H_\nu^{(j)}(a|x|) \big] \big| \le C_{85} |a|^{k - \frac{1}{2}} |x|^{-\frac{1}{2}} \mathrm{e}^{(-1)^j |x| \,\mathrm{Im}\, a} \qquad (A.71)$$

for all $x \in \mathbb{R}^n$ with $|x| \ge \varepsilon$ and $a \in \mathbb{C}$ with $|a| \ge \delta$.

Proof. For $|a| \ge \delta$ and $|x| \ge \varepsilon$ we have $|a|x|| \ge \delta\varepsilon > 0$. Therefore, by (A.67) we can estimate

$$\big| H_\nu^{(j)}(a|x|) \big| \le c_0 |a|x||^{-\frac{1}{2}} \mathrm{e}^{(-1)^j |x| \,\mathrm{Im}\, a} .$$

From equation (A.70) we now obtain

$$\big| \nabla^k \big[H_\nu^{(j)}(a|x|) \big] \big| \le C_{85} |x|^{-k} \big(|a|\,|x| \big)^{-\frac{1}{2}} \mathrm{e}^{(-1)^j |x| \,\mathrm{Im}\, a} \sum_{\ell=0}^{k} \big(|a|\,|x| \big)^\ell .$$

Due to $|a|\,|x| \ge \delta\varepsilon$, the final sum is bounded by a constant multiple of $\big(|a|\,|x| \big)^k$. This shows (A.71). $\qquad\square$

A.1.2 A Function with Anisotropic Decay

This section is dedicated to the study of the function

$$g_{a,b} \colon \mathbb{R}^n \setminus \{0\} \to \mathbb{R}, \qquad g_{a,b}(x) := |x|^a \, e^{bx_1} \tag{A.72}$$

for parameters $a, b \in \mathbb{R}$ and $n \in \mathbb{N}$. To derive a formula for the derivatives of $g_{a,b}$, we further define the function $h = (h_1, \ldots, h_n)$ by

$$h \colon \mathbb{R}^n \setminus \{0\} \to \mathbb{R}^n, \qquad h(x) = \frac{ax}{|x|^2} + b\,e_1 .$$

Then a simple calculation shows

$$\partial_j g_{a,b}(x) = a|x|^{a-2} x_j \, e^{bx_1} + |x|^a b \delta_{1j} \, e^{bx_1} = h_j(x) g_{a,b}(x). \tag{A.73}$$

For higher derivatives we obtain the following representation.

Lemma A.1.4. *Let $a, b \in \mathbb{R}$, $n \in \mathbb{N}$ and $\alpha \in \mathbb{N}_0^n$. Then*

$$D^\alpha g_{a,b}(x) = \sum_{\beta \le \alpha} \frac{p_\beta(x)}{|x|^{2|\beta|}} h^{\alpha-\beta}(x) g_{a,b}(x) \tag{A.74}$$

for all $x \in \mathbb{R}^n \setminus \{0\}$, where $p_\beta \colon \mathbb{R}^n \to \mathbb{R}$ are polynomials in x such that $\deg p_{\alpha,\ell} \le |\beta|$. Here, $h^\gamma = h_1^{\gamma_1} \cdots h_n^{\gamma_n}$.

Proof. We show the statement inductively. For $\alpha = 0$, there is nothing to do. So assume that representation (A.74) holds for some $\alpha \in \mathbb{N}_0^n$. Then with (A.73) we compute

$$\partial_j D^\alpha g_{a,b}(x)$$

$$= \sum_{\beta \le \alpha} \left[\frac{\partial_j p_\beta(x)}{|x|^{2|\beta|}} h^{\alpha-\beta}(x) g_{a,b}(x) + \frac{-2|\beta| x_j p_\beta(x)}{|x|^{2|\beta|+2}} h^{\alpha-\beta}(x) g_{a,b}(x) \right.$$

$$\left. + \frac{p_\beta(x)}{|x|^{2|\beta|}} \left(\sum_{k=1}^n (\alpha_k - \beta_k) h^{\alpha-\beta-e_k}(x) \partial_j h_k(x) g_{a,b}(x) + h^{\alpha-\beta}(x) \partial_j g_{a,b}(x) \right) \right]$$

$$= \sum_{e_j \le \beta \le \alpha+e_j} \frac{\partial_j p_{\beta-e_j}(x) |x|^2 - 2(|\beta|-1) x_j p_{\beta-e_j}(x)}{|x|^{2|\beta|}} h^{\alpha+e_j-\beta}(x) g_{a,b}(x)$$

$$+ \sum_{k=1}^n \sum_{\substack{\beta \ge e_j + e_k \\ \beta \le \alpha + e_j + e_k}} (\alpha_k - \beta_k + \delta_{jk} + 1) \frac{p_{\beta-e_j-e_k}(x)}{|x|^{2|\beta|-4}} \partial_j h_k(x) h^{\alpha+e_j-\beta}(x) g_{a,b}(x)$$

$$+ \sum_{\beta \le \alpha} \frac{p_\beta(x)}{|x|^{2|\beta|}} h^{\alpha+e_j-\beta}(x) g_{a,b}(x).$$

Using now the identity

$$\partial_j h_k(x) = \frac{a|x|^2 \delta_{jk} - 2ax_j x_k}{|x|^4},$$

we see that $\partial_j D^\alpha g_{a,b}(x)$ has the desired form, and the assertion follows by induction. $\qquad\Box$

Identity (A.74) gives rise to the following estimate.

Lemma A.1.5. *Let $a, b \in \mathbb{R}$, $n \in \mathbb{N}$, $k \in \mathbb{N}_0$ and $\varepsilon > 0$. Then there exists a constant $C_{86} = C_{86}(n, k, a, b, \varepsilon) > 0$ such that*

$$\left|\nabla^k g_{a,b}(x)\right| \le C_{86}|x|^a \, e^{bx_1} \tag{A.75}$$

for all $x \in \mathbb{R}^n$ with $|x| \ge \varepsilon$.

Proof. Since $|h(x)| \le c_0$ for $|x| \ge \varepsilon$, formula (A.74) implies

$$\left|\nabla^k g_{a,b}(x)\right| \le c_1 \sum_{\ell=0}^{k} |x|^{-\ell} |g_{a,b}(x)| \le c_2 |g_{a,b}(x)| \le c_3 |x|^a \, e^{bx_1}$$

for $|x| \ge \varepsilon$. This shows the statement. $\qquad\Box$

A.2 Convolutions

A.2.1 Derivatives of Convolutions

Lemma A.2.1. *Let $f, g \in \mathrm{L}^1_{\mathrm{loc}}(\mathbb{R}^n) \cap \mathrm{C}^0(\mathbb{R}^n \setminus \{0\})$, and assume that there exists $R > 0$ and $\alpha, \beta > 0$ with $\alpha + \beta > n$ such that*

$$\sup_{x \in B^R} |x|^\alpha |f(x)| + |x|^\beta |g(x)| < \infty. \tag{A.76}$$

Then the convolution integral

$$f * g(x) = \int_{\mathbb{R}^n} f(x - y) g(y) \, \mathrm{d}y$$

is well defined for $x \ne 0$ and $f * g \in \mathrm{L}^1_{\mathrm{loc}}(\mathbb{R}^n)$. If additionally $g \in \mathrm{W}^{1,1}_{\mathrm{loc}}(\mathbb{R}^n) \cap \mathrm{C}^1(\mathbb{R}^n \setminus \{0\})$ and there exists $\gamma > 0$ with $\alpha + \gamma > n$ such that

$$\sup_{x \in B^R} |x|^\gamma |\nabla g(x)| < \infty, \tag{A.77}$$

then $f * g \in \mathrm{C}^1(\mathbb{R}^n \setminus \{0\})$ and

$$\partial_j (f * g)(x) = (f * \partial_j g)(x) = \int_{\mathbb{R}^n} f(x - y) \partial_j g(y) \, \mathrm{d}y \qquad (j = 1, \dots, n).$$

Proof. First note that by continuity of f and g in $\mathbb{R}^n \setminus \{0\}$, property (A.76) holds for all $R > 0$. For $x \neq 0$ we set $R = |x|/2$. Then both f and g are continuous and thus bounded on the closure of $B_R(x)$, which yields

$$\int_{B_R(0)} |f(x-y)||g(y)|\,dy \leq \sup_{z \in B_R(x)} |f(z)| \int_{B_R(0)} |g(y)|\,dy < \infty,$$

$$\int_{B_R(x)} |f(x-y)||g(y)|\,dy \leq \sup_{y \in B_R(x)} |g(y)| \int_{B_R(0)} |f(z)|\,dz < \infty.$$

Moreover, because $y \in B^R(0) \cap B^R(x)$ implies $|y| \leq |x-y| + |x| \leq 3|x-y|$, from (A.76) we deduce

$$\int_{B^R(0) \cap B^R(x)} |f(x-y)||g(y)|\,dy \leq c_0 \int_{B^R(0) \cap B^R(x)} |x-y|^{-\alpha}|y|^{-\beta}\,dy$$

$$\leq c_1 \int_{B^R(0) \cap B^R(x)} |y|^{-\alpha-\beta}\,dy < \infty$$

since $\alpha + \beta > n$. Collecting these integrals, we conclude

$$\int_{\mathbb{R}^n} |f(x-y)||g(y)|\,dy < \infty,$$

so that the convolution integral $f * g(x)$ exists for all $x \neq 0$. Now let $R_0 > 0$ be arbitrary. Then we have

$$\int_{B_{R_0}} |f * g(x)|\,dx \leq \int_{B_{R_0}} \int_{\mathbb{R}^n} |f(x-y)||g(y)|\,dy dx$$

$$= \int_{B_{2R_0}} \int_{B_{R_0}} |f(x-y)||g(y)|\,dx dy + \int_{B^{2R_0}} \int_{B_{R_0}} |f(x-y)||g(y)|\,dx dy.$$

Since $|x| \leq R_0$ and $|y| \leq 2R_0$ implies $|x-y| \leq 3R_0$, the first integral can be estimated by

$$\int_{B_{2R_0}} \int_{B_{R_0}} |f(x-y)||g(y)|\,dx dy \leq \int_{B_{3R_0}} |f(z)|\,dz \int_{B_{2R_0}} |g(y)|\,dy < \infty.$$

Moreover, since $|x| \leq R_0$ and $|y| \geq 2R_0$ implies $|x-y| \geq |y| - |x| \geq |y| - R_0 \geq |y|/2 \geq R_0$, we can employ (A.76) to estimate the second integral by

$$\int_{B^{2R_0}} \int_{B_{R_0}} |f(x-y)||g(y)|\,dx dy \leq c_2 \int_{B^{2R_0}} \int_{B_{R_0}} |x-y|^{-\alpha}|y|^{-\beta}\,dx dy$$

$$\leq c_3 \int_{B^{2R_0}} \int_{B_{R_0}} |y|^{-\alpha}|y|^{-\beta}\,dx dy = c_3 |B_{R_0}| \int_{B^{2R_0}} |y|^{-\alpha-\beta}\,dy < \infty.$$

which follows from $\alpha + \beta > n$. In total, this shows $f * g \in L^1_{loc}(\mathbb{R}^n)$.

Now assume $g \in W^{1,1}_{loc}(\mathbb{R}^n) \cap C^1(\mathbb{R}^n \setminus \{0\})$ and (A.77). With exactly the same argument as above we see that the convolution integral $f * \partial_j g(x)$ exists for $x \neq 0$. Now let $\varepsilon, \rho > 0$ and $h \in \mathbb{R} \setminus \{0\}$, and consider

$$\frac{1}{h}\big(f * g(x + h\,e_j) - f * g(x)\big) - f * \partial_j g(x)$$

$$= \int_{\mathbb{R}^n} f(x - y)\left[\frac{1}{h}\big(g(y + h\,e_j) - g(y)\big) - \partial_j g(y)\right] dy = I_1 + I_2 + I_3$$

for $x \neq 0$, where we set

$$I_1 := \int_{B_\delta} f(x - y)\left[\int_0^1 \partial_j g(y + s h\,e_j)\,ds - \partial_j g(y)\right] dy,$$

$$I_2 := \int_{B^\rho} f(x - y)\left[\int_0^1 \partial_j g(y + s h\,e_j)\,ds - \partial_j g(y)\right] dy,$$

$$I_3 := \int_{B_\rho \setminus B_\delta} f(x - y)\left[\int_0^1 \partial_j g(y + s h\,e_j)\,ds - \partial_j g(y)\right] dy.$$

For $\delta \leq R = |x|/2$ we obtain

$$|I_1| \leq \sup_{z \in B^R} |f(z)| \left(\int_{B_{\delta + |h|}} |\partial_j g(y)|\,dy + \int_{B_\delta} |\partial_j g(y)|\,dy \right)$$

$$\leq 2 \sup_{z \in B^R} |f(z)| \int_{B_{\delta + |h|}} |\partial_j g(y)|\,dy.$$

If we choose δ sufficiently small and $|h| \leq \delta$, we thus have $|I_1| \leq \varepsilon/3$. Moreover, for $\rho \geq 4R = 2|x|$ and $|h| \leq \rho/2$ we exploit (A.76) and (A.77) and utilize that $|y| \geq \rho$ implies $|x - y| \geq |y| - |x| \geq |y|/2$ to estimate

$$|I_2| \leq c_4 \int_0^1 \int_{B^\rho} |x - y|^{-\alpha}|y + s h\,e_j|^{-\gamma}\,ds dy + \int_{B^\rho} |x - y|^{-\alpha}|y|^{-\gamma}\,dy$$

$$\leq c_5 \int_{B^{\rho - |h|}} |y|^{-\alpha - \gamma}\,dy.$$

Therefore, we obtain $|I_2| \leq \varepsilon/3$ for ρ sufficiently large. Furthermore, due to $f(x - \cdot) \in L^1(B_\rho \setminus B_\delta)$ and $\partial_j g \in C(B_\rho \setminus B_\delta)$ with $\rho < \delta$ as above, the dominated convergence theorem yields $\lim_{|h|\to\infty} I_3 = 0$, that is, $|I_3| \le \varepsilon/3$ for $|h|$ sufficiently small. In total, we obtain

$$\left|\frac{1}{h}\big(f * g(x + h\,e_j) - f * g(x)\big) - f * \partial_j g(x)\right| \le \varepsilon$$

for $|h|$ sufficiently small, which implies $\partial_j(f * g) = f * \partial_j g$ in $\mathbb{R}^n \setminus \{0\}$. By the same argument as above, we further obtain $\partial_j(f * g) \in L^1_{\mathrm{loc}}(\mathbb{R}^n)$, which completes the proof. $\qquad\square$

A.2.2 Estimates of Convolutions

The following convolution estimate treats functions with decay estimates that include the anisotropic function $s(x) := |x| + x_1$. As establishing such kind of estimates is a cumbersome work, we do not give a proof here.

Lemma A.2.2. Let $A \in (2, \infty)$, $B \in [0, \infty)$ with $A + \min\{1, B\} > 3$. Then there exists $C_{87} = C_{87}(A, B) > 0$ such that for all $x \in \mathbb{R}^3$ it holds

$$\int_{\mathbb{R}^3} \big[(1 + |x - y|)(1 + s(x - y))\big]^{-3/2}(1 + |y|)^{-A}(1 + s(y))^{-B}\, dy$$

$$\le C_{87}(1 + |x|)^{-3/2}.$$

Proof. We refer to [71, Proof of Theorem 3.2]. $\qquad\square$

We also need the following lemma treating convolutions of homogeneous functions.

Lemma A.2.3. Let $A \in (0, 3)$, $B \in (0, \infty)$, $\alpha \in (0, \infty)$. Then there exists a constant $C_{88} = C_{88}(A, B, \alpha) > 0$ such that for all $x \in \mathbb{R}^3 \setminus \{0\}$ it holds

$$\int_{\mathbb{R}^3} |x - y|^{-A}\, e^{-\alpha|x-y|}(1 + |y|)^{-B}\, dy \le C_{88}|x|^{-B}.$$

Proof. We split the integral into two parts

$$I_1 := \int_{B_{|x|/2}(x)} |x - y|^{-A}\, e^{-\alpha|x-y|}(1 + |y|)^{-B}\, dy,$$

$$I_2 := \int_{B^{|x|/2}(x)} |x - y|^{-A}\, e^{-\alpha|x-y|}(1 + |y|)^{-B}\, dy,$$

which we estimate separately. On the one hand, since $|x - y| \leq |x|/2$ implies $|y| \geq |x| - |x - y| \geq |x|/2$, we have

$$I_1 \leq c_0(1 + |x|)^{-B} \int_{\mathbb{R}^3} |x - y|^{-A} e^{-\alpha|x-y|} \, dy \leq c_1(1 + |x|)^{-B} \leq c_2|x|^{-B},$$

where the integral is finite due to $A < 3$. On the other hand, for the second integral we directly obtain

$$I_2 \leq c_3 e^{-\alpha|x|/4} \int_{\mathbb{R}^3} e^{-\alpha|x-y|/2} \, dy \leq c_4 e^{-\alpha|x|/4} \leq c_5|x|^{-B}$$

for all $x \neq 0$. This completes the proof. $\qquad\square$

A.3 Classical Fourier Analysis

A.3.1 Fourier Transforms of Elementary Functions

Here we provide properties of Fourier transforms of specific functions that are occasionally encountered in the course of this thesis.

Proposition A.3.1. *For $\alpha \in (0, 1)$ and $s \in [1, \infty)$ define*

$$\varphi_\alpha := \mathscr{F}_{\mathbb{T}}^{-1}\big[k \mapsto (1 - \delta_{\mathbb{Z}}(k))|k|^{-\alpha}\big]. \tag{A.78}$$

Then $\varphi_\alpha \in L^{s,\infty}(\mathbb{T})$ if $s \leq 1/(1 - \alpha)$, and $\varphi_\alpha \in L^s(\mathbb{T})$ if $s < 1/(1 - \alpha)$.

Proof. We choose $(-\frac{T}{2}, \frac{T}{2}]$ as a representation of $\mathbb{T} = \mathbb{R}/T\mathbb{Z}$. Then we have $\varphi_\alpha(t) = c_0|t|^{\alpha-1} + h_\alpha(t)$ for some function $h_\alpha \in C^\infty(\mathbb{T})$; see [56, Example 3.1.19] for example. This directly yields the claim. $\qquad\square$

Proposition A.3.2. *Let $n \in \mathbb{N}$. For $\beta \in (0, n)$ and $s \in [1, \infty)$ define*

$$\psi_\beta := \mathscr{F}_{\mathbb{R}^n}^{-1}\big[\xi \mapsto (1 + |\xi|^2)^{-\beta/2}\big]. \tag{A.79}$$

Then $\psi_\beta \in L^{s,\infty}(\mathbb{R}^n)$ if $s \leq n/(n - \beta)$, and $\psi_\beta \in L^s(\mathbb{R}^n)$ if $s < n/(n - \beta)$.

Proof. We have

$$0 < \psi_\beta(x) \leq c_0 \chi_{(0,2)}(|x|)|x|^{\beta-n} + c_1 \chi_{[2,\infty)}(|x|) e^{-|x|/2};$$

see [57, Proposition 6.1.5] for example. This directly yields the claim. $\qquad\square$

A.3.2 The Marcinkiewicz Multiplier Theorem

There are several theorems, which nowadays belong to the standard repertoire in Fourier analysis and give sufficient conditions for a function m to be an L^p multiplier in the Euclidean setting $G = \mathbb{R}^n$. One of these is the Marcinkiewicz Multiplier Theorem, which can be applied to functions that are sufficiently regular away from the coordinate axes in $\mathbb{R}^n$, that is, on the set

$$\mathbb{R}_c^n := \left\{ x = (x_1, \dots, x_n) \in \mathbb{R}^n \mid x_j \neq 0 \text{ for at least two } j \in \{1, \dots, n\} \right\}.$$

Theorem A.3.3 (Marcinkiewicz Multiplier Thereom). *Let $n \in \mathbb{N}$, and let $m \in C^n(\mathbb{R}_c^n)$ be a bounded function such that*

$$A := \sup_{\alpha \in \{0,1\}^n} \sup_{\xi \in \mathbb{R}_c^n} \left| \xi^\alpha D_\xi^\alpha m(\xi) \right| < \infty. \tag{A.80}$$

Then m is an $L^p(\mathbb{R}^n)$ multiplier for any $p \in (1, \infty)$, and

$$\left\| \mathrm{op}_{\mathbb{R}^n}[m] \right\|_{\mathcal{L}(L^p(\mathbb{R}^n))} \leq C_{89} A$$

for some constant $C_{89} = C_{89}(n, p) > 0$.

Proof. A proof can be found in [56, Corollary 5.2.5] for example. □

A simple consequence is the L^p continuity of the Riesz transforms.

Proposition A.3.4 (Riesz Transform). *For $j \in \{1, \dots, n\}$ let $\mathfrak{R}_j$ denote the* Riesz transform *given by*

$$\mathfrak{R}_j \colon \mathscr{S}(\mathbb{R}^n) \to \mathscr{S}'(\mathbb{R}^n), \quad \mathfrak{R}_j(f) := \mathscr{F}_{\mathbb{R}^n}^{-1}\left[\frac{-i\xi_j}{|\xi|} \mathscr{F}_{\mathbb{R}^n}(f) \right]. \tag{A.81}$$

Then $\mathfrak{R}_j$ can be extended to a continuous linear operator $\mathfrak{R}_j \in \mathcal{L}(L^p(\mathbb{R}^n))$ for any $p \in (1, \infty)$.

Proof. This is a direct consequence of Theorem A.3.3. For a different proof see [56, Corollary 4.2.8] for example. □

A.3.3 Some Multipliers

Here we study specific functions and there properties as Fourier multipliers. To begin with, let $\chi \in C_0^\infty(\mathbb{R}; [0, 1])$ be a cut-off function with

$$\chi(\eta) = 1 \quad \text{for } |\eta| \leq \frac{1}{2}, \qquad \chi(\eta) = 0 \quad \text{for } |\eta| \geq 1. \tag{A.82}$$

Let $\lambda \in \mathbb{R}$ and $\mathcal{T} > 0$. For $\theta \in [0,1]$ and $h = 1, \ldots, n$ we define

$$m_0 \colon \mathbb{R} \times \mathbb{R}^n \to \mathbb{C}, \quad m_0(\eta, \xi) := \frac{(1 - \chi(\eta))|\eta|^\theta \left(1 + |\xi|^2\right)^{1-\theta}}{|\xi|^2 + i\left(\frac{2\pi}{\mathcal{T}}\eta - \lambda\xi_1\right)}, \tag{A.83}$$

$$m_h \colon \mathbb{R} \times \mathbb{R}^n \to \mathbb{C}, \quad m_h(\eta, \xi) := \frac{(1 - \chi(\eta))|\eta|^\theta \left(1 + |\xi|^2\right)^{\frac{1}{2}-\theta} i\xi_h}{|\xi|^2 + i\left(\frac{2\pi}{\mathcal{T}}\eta - \lambda\xi_1\right)}. \tag{A.84}$$

In order to show that these functions really define $L^q(\mathbb{R} \times \mathbb{R}^n)$ multipliers, we first give a lower bound of the denominator

$$N_\lambda(\eta, \xi) := |\xi|^2 + i\left(\frac{2\pi}{\mathcal{T}}\eta - \lambda\xi_1\right).$$

Lemma A.3.5. *There is a constant $C_{90} = C_{90}(\mathcal{T}, \lambda) > 0$ such that*

$$|N_\lambda(\eta, \xi)| \ge C_{90}\left(1 + |\eta| + |\xi|^2\right) \tag{A.85}$$

for all $(\eta, \xi) \in \mathbb{R} \times \mathbb{R}^n$ with $|\eta| \ge \frac{1}{2}$.

Proof. First note that we trivially have $|N_\lambda(\eta, \xi)| \ge |\xi|^2$. Moreover, for $\lambda = 0$ we obtain $|N_\lambda(\eta, \xi)| \ge \frac{2\pi}{\mathcal{T}}|\eta| \ge \frac{\pi}{2\mathcal{T}}(1 + 2|\eta|)$, so that (A.85) follows immediately in this case. Now consider $\lambda \ne 0$. For $|\lambda||\xi| \le \frac{\pi}{\mathcal{T}}|\eta|$ we have

$$|N_\lambda(\eta, \xi)| \ge \left|\frac{2\pi}{\mathcal{T}}\eta - \lambda\xi_1\right| \ge \frac{2\pi}{\mathcal{T}}|\eta| - |\lambda||\xi| \ge \frac{\pi}{\mathcal{T}}|\eta| \ge \frac{\pi}{4\mathcal{T}}(1 + 2|\eta|),$$

and for $|\lambda||\xi| \ge \frac{\pi}{\mathcal{T}}|\eta|$ we obtain

$$|N_\lambda(\eta, \xi)| \ge |\xi|^2 \ge \frac{\pi^2}{\lambda^2\mathcal{T}^2}|\eta|^2 \ge \frac{\pi^2}{2\lambda^2\mathcal{T}^2}|\eta| \ge \frac{\pi^2}{8\lambda^2\mathcal{T}^2}(1 + 2|\eta|).$$

Combining these estimates with $|N_\lambda(\eta, \xi)| \ge |\xi|^2$, we also conclude (A.85) in the case $\lambda \ne 0$. $\qquad\square$

Next, we derive a representation formula for the derivatives of m_h with respect to ξ.

Lemma A.3.6. *Let $h \in \{0, \ldots, n\}$ and $\alpha \in \mathbb{N}_0^n$. Then $m_h \in C^\infty(\mathbb{R} \times \mathbb{R}^n)$ and*

$$D_\xi^\alpha m_h(\eta, \xi) = (1 - \chi(\eta))|\eta|^\theta \sum_{\ell=0}^{|\alpha|} \frac{p_{\alpha,\ell}(\xi)}{N_\lambda(\eta, \xi)^{\ell+1}(1 + |\xi|^2)^{|\alpha|-\ell-\zeta}}, \tag{A.86}$$

where $p_{\alpha,\ell} \colon \mathbb{R}^n \to \mathbb{C}$ are complex-valued polynomials, and

$$\deg p_{\alpha,\ell} \le \begin{cases} |\alpha| & \text{if } h = 0, \\ |\alpha| + 1 & \text{if } h \in \{1, \ldots, n\}, \end{cases} \qquad \zeta = \begin{cases} 1 - \theta & \text{if } h = 0, \\ \frac{1}{2} - \theta & \text{if } h \in \{1, \ldots, n\}. \end{cases}$$

Proof. By Lemma A.3.5, the denominator $N_\lambda(\eta, \xi)$ is bounded from below for $|\eta| \geq \frac{1}{2}$, and $m_h(\eta, \xi) = 0$ for $|\eta| \leq \frac{1}{2}$. Since all involved functions are smooth, we thus conclude that m_h is smooth on $\mathbb{R} \times \mathbb{R}^n$. We show formula (A.86) inductively. For $\alpha = 0$ there is nothing to show. Hence assume that (A.86) holds for some $\alpha \in \mathbb{N}_0^n$. Differentiating the terms of the sum in (A.86) separately, we obtain

$$\partial_{\xi_j} \left[\frac{p_{\alpha,\ell}(\xi)}{N_\lambda(\eta, \xi)^{\ell+1}(1 + |\xi|^2)^{|\alpha|-\ell-\varsigma}} \right] = \frac{\partial_j p_{\alpha,\ell}(\xi)(1 + |\xi|^2)}{N_\lambda(\eta, \xi)^{\ell+1}(1 + |\xi|^2)^{|\alpha|+1-\ell-\varsigma}}$$
$$- \frac{p_{\alpha,\ell}(\xi)(\ell+1)(2\xi_j - i\lambda\delta_{1j})}{N_\lambda(\eta, \xi)^{\ell+2}(1 + |\xi|^2)^{|\alpha|-\ell-\varsigma}} - \frac{p_{\alpha,k}(\xi)2(|\alpha| - \ell - \varsigma)\xi_j}{N_\lambda(\eta, \xi)^{\ell+1}(1 + |\xi|^2)^{|\alpha|+1-\ell-\varsigma}}.$$

Therefore, $\partial_{\xi_j} D_\xi^\alpha m_h$ is also of the asserted structure. This completes the proof. $\qquad \square$

These preparations enable us to show that m_h is an $L^q(\mathbb{R} \times \mathbb{R}^n)$ multiplier by employing the Marcinkiewicz Multiplier Theorem (Theorem A.3.3).

Lemma A.3.7. *The function $m_h \in C^\infty(\mathbb{R} \times \mathbb{R}^n)$ is an $L^q(\mathbb{R} \times \mathbb{R}^n)$ multiplier for all $q \in (1, \infty)$ and $h \in \{0, \ldots, n\}$.*

Proof. By Lemma A.3.6, m_h is smooth. Moreover, by (A.85) and (A.86), we obtain

$$\left| D_\xi^\alpha m_h(\eta, \xi) \right| \leq \sum_{\ell=0}^{|\alpha|} \frac{|\eta|^\theta (1 + |\xi|^2)^\varsigma |p_{\alpha,\ell}(\xi)|}{(1 + |\eta| + |\xi|^2)(1 + |\xi|^2)^{|\alpha|}}.$$

If $h = 0$, we have $\theta + \eta = 1$, so that Young's inequality and $\deg p_{\alpha,k} \leq |\alpha|$ imply

$$\left| \xi^\alpha D_\xi^\alpha m_h(\eta, \xi) \right| \leq c_0 |\xi|^{|\alpha|} \frac{(|\eta| + 1 + |\xi|^2)(1 + |\xi|^2)^{|\alpha|/2}}{(1 + |\eta| + |\xi|^2)(1 + |\xi|^2)^{|\alpha|}} \leq c_0.$$

If $h \in \{1, \ldots, n\}$, we have $\theta + \varsigma + \frac{1}{2} = 1$, and Young's inequality and $\deg p_{\alpha,\ell} \leq |\alpha| + 1$ lead to the very same estimate. Next we compute the derivative with respect to η. Identity (A.86) yields

$$\partial_\eta D_\xi^\alpha m_h(\eta, \xi)$$

$$= \left((1 - \chi(\eta))\theta |\eta|^{\theta-2}\eta + \chi'(\eta)|\eta|^\theta \right) \sum_{\ell=0}^{|\alpha|} \frac{p_{\alpha,\ell}(\xi)}{N_\lambda(\eta, \xi)^{\ell+1}(1 + |\xi|^2)^{|\alpha|-\ell-\varsigma}}$$

$$- (1 - \chi(\eta))|\eta|^\theta \sum_{\ell=0}^{|\alpha|} \frac{p_{\alpha,\ell}(\xi)(\ell+1)i\frac{2\pi}{T}}{N_\lambda(\eta, \xi)^{\ell+2}(1 + |\xi|^2)^{|\alpha|-\ell-\varsigma}}.$$

Employing (A.85), we thus have

$$|\eta\xi^\alpha\partial_\eta D_\xi^\alpha m_h(\eta,\xi)| \le c_1\left(1 + \frac{|\eta|}{1+|\eta|+|\xi|^2}\right)|\xi|^{|\alpha|}\sum_{\ell=0}^{|\alpha|}\frac{|\eta|^\theta(1+|\xi|^2)^\varsigma|p_{\alpha,\ell}(\xi)|}{(1+|\eta|+|\xi|^2)(1+|\xi|^2)^{|\alpha|}}.$$

With the same argument as above, we see that the remaining terms are bounded. In particular, we conclude

$$\sup\left\{|\eta^\beta\xi^\alpha\partial_\eta^\beta D_\xi^\alpha m_h(\eta,\xi)| \,\middle|\, \alpha \in \{0,1\}^n,\ \beta \in \{0,1\},\ (\eta,\xi) \in \mathbb{R}\times\mathbb{R}^n\right\} < \infty.$$

Now the Marcinkiewicz Multiplier Theorem (Theorem A.3.3) implies that m_h is an $L^q(\mathbb{R}\times\mathbb{R}^n)$ multiplier for all $q \in (1,\infty)$. $\qquad\square$

For $\theta \in [0,1]$, $\kappa > 0$ and $\lambda \in \mathbb{R}$, we next consider the function

$$\widetilde{m}_{\kappa,\lambda}\colon\mathbb{R}\times\mathbb{R}^n \to \mathbb{C},\quad \widetilde{m}_{\kappa,\lambda}(\eta,\xi) := \frac{(1-\chi(\eta))|\kappa\eta|^\theta|\xi|^{2-2\theta}}{|\xi|^2 + i(\kappa\eta - \lambda\xi_1)} \tag{A.87}$$

for a cut-off function $\chi \in C_0^\infty(\mathbb{R};[0,1])$ satisfying (A.82) as above. Observe that $\widetilde{m}_{\kappa,\lambda}$ can be seen as a homogeneous version of (A.83) with $\lambda = 0$. Similarly to Lemma A.3.6, we obtain the following representation formula for derivatives of $\widetilde{m}_{\kappa,\lambda}$. Since we are interested in the dependencies on the parameters κ and λ, we only consider derivatives with respect to $\xi_2, \cdots, \xi_n$ at first. We set

$$N_{\kappa,\lambda}(\xi,\eta) := |\xi|^2 + i(\kappa\eta - \lambda\xi_1).$$

Lemma A.3.8. *The function $\widetilde{m}_{\kappa,\lambda}$ is continuous on $\mathbb{R}\times\mathbb{R}^n$ and smooth on the set $\{(\eta,\xi) \in \mathbb{R}\times\mathbb{R}^n \mid |\xi| \ne 0\}$ with*

$$D_\xi^\alpha\widetilde{m}_{\kappa,\lambda}(\eta,\xi) = (1-\chi(\eta))|\kappa\eta|^\theta\sum_{\ell=0}^{|\alpha|}\frac{p_{\alpha,\ell}(\xi)}{N_{\kappa,\lambda}(\xi,\eta)^{\ell+1}|\xi|^{2|\alpha|-2\ell-2+2\theta}} \tag{A.88}$$

for all $\alpha \in \mathbb{N}_0^n$ with $\alpha_1 = 0$, where $p_{\alpha,\ell}\colon\mathbb{R}^n \to \mathbb{C}$ are complex-valued homogeneous polynomials of degree $\deg p_{\alpha,\ell} = |\alpha|$ that are independent of κ and λ, or $p_{\alpha,\ell} \equiv 0$.

Proof. The proof works analogous to that of Lemma A.3.6. First of all, the denominator in (A.87) is bounded from below for $|\eta| \ge \frac{1}{2}$, and $\widetilde{m}_{\kappa,\lambda}(\eta,\xi) = 0$ for $|\eta| \le \frac{1}{2}$. Since all involved functions are smooth for $|\xi| \ne 0$, we thus conclude that $\widetilde{m}_{\kappa,\lambda}$ is smooth for $|\xi| \ne 0$, and $\widetilde{m}_{\kappa,\lambda}$ is continuous on $\mathbb{R}\times\mathbb{R}^n$. We prove formula (A.88) inductively. For $\alpha = 0$ there is nothing to show.

Hence assume that (A.88) holds for some $\alpha \in \mathbb{N}_0^n$. Differentiating the terms of the sum in (A.88) separately, we obtain

$$
\partial_{\xi_j} \left[\frac{p_{\alpha,\ell}(\xi)}{N_{\kappa,\lambda}(\xi,\eta)^{\ell+1}|\xi|^{2|\alpha|-2\ell-2+2\theta}} \right] = \frac{\partial_j p_{\alpha,\ell}(\xi)|\xi|^2}{N_{\kappa,\lambda}(\xi,\eta)^{\ell+1}|\xi|^{2|\alpha|+2-2\ell-2+2\theta}}
$$
$$
- \frac{p_{\alpha,\ell}(\xi)(\ell+1)2\xi_j}{N_{\kappa,\lambda}(\xi,\eta)^{\ell+2}|\xi|^{2|\alpha|-2\ell-2+2\theta}} - \frac{p_{\alpha,\ell}(\xi)(2|\alpha|-2\ell-2+2\theta)\xi_j}{N_{\kappa,\lambda}(\xi,\eta)^{\ell+1}|\xi|^{2|\alpha|+2-2\ell-2+2\theta}}.
$$

Therefore, $\partial_{\xi_j} D_\xi^\alpha \widetilde{m}_{\kappa,\lambda}$ is also of the asserted structure. This completes the proof. $\qquad\square$

In order to show that $\widetilde{m}_{\kappa,\lambda}$ is an L^q multiplier, we proceed as in Lemma A.3.7 and employ the Marcinkiewicz Multiplier Theorem (Theorem A.3.3). To this end, we prepare the following estimates

Lemma A.3.9. *There exists a polynomial* $P\colon\mathbb{R}\to\mathbb{R}$ *such that*

$$
\frac{|\xi|^2 + |\lambda\xi_1| + |\kappa\eta|}{|N_{\kappa,\lambda}(\xi,\eta)|} \le P(\lambda^2/\kappa) \tag{A.89}
$$

for all $|\eta| \ge 1/2$.

Proof. Clearly, we have

$$
\frac{|\xi|^2}{|N_{\kappa,\lambda}(\xi,\eta)|} \le 1.
$$

For the remaining estimates, we proceed as in Lemma A.3.5 and distinguish two cases. If $|\lambda\xi| \le \kappa|\eta|/2$, we have $|\kappa\eta - \lambda\xi_1| \ge \kappa|\eta| - |\lambda||\xi| \ge \kappa|\eta|/2 \ge |\lambda||\xi|$, so that

$$
\frac{|\lambda\xi_1| + |\kappa\eta|}{|N_{\kappa,\lambda}(\xi,\eta)|} \le \frac{3|\kappa\eta - \lambda\xi_1|}{|\kappa\eta - \lambda\xi_1|} \le 3.
$$

If $|\lambda\xi| > \kappa|\eta|/2$, we obtain $|\lambda\xi| \ge \kappa/4$ and thus

$$
\frac{|\lambda\xi_1| + |\kappa\eta|}{|N_{\kappa,\lambda}(\xi,\eta)|} \le \frac{3|\lambda||\xi|}{|\xi|^2} \le \frac{12\lambda^2}{\kappa}.
$$

In total, this shows (A.89). $\qquad\square$

With estimate (A.89) at hand, we can now prove that $\widetilde{m}_{\kappa,\lambda}$ is an L^q multiplier and we obtain a bound on the multiplier norm that is uniform in λ and ω as long as $\lambda^2 \le \theta\omega$ for some $\theta > 0$.

Lemma A.3.10. *The function $\widetilde{m}_{\kappa,\lambda} \in C(\mathbb{R} \times \mathbb{R}^n)$ is an $L^q(\mathbb{R} \times \mathbb{R}^n)$ multiplier for all $q \in (1,\infty)$, and there exists a polynomial $P\colon \mathbb{R} \to \mathbb{R}$ such that*

$$\|\mathrm{op}_{\mathbb{R} \times \mathbb{R}^n}[\widetilde{m}_{\kappa,\lambda}]\|_{\mathcal{L}(L^q(\mathbb{R} \times \mathbb{R}^n))} \leq P(\lambda^2/\kappa). \tag{A.90}$$

Proof. By Lemma A.3.8, $\widetilde{m}_{\kappa,\lambda}$ is continuous, and (A.88) yields

$$\left| D_\xi^\alpha \widetilde{m}_{\kappa,\lambda}(\eta,\xi) \right| \leq \sum_{\ell=0}^{|\alpha|} \frac{|\kappa\eta|^\theta |\xi|^{2-2\theta} |p_{\alpha,\ell}(\xi)|}{\left|N_{\kappa,\lambda}(\xi,\eta)\right|^{\ell+1} |\xi|^{2|\alpha|-2\ell}}$$

for $\alpha \in \mathbb{N}_0^n$ with $\alpha_1 = 0$. From the homogeneity of $p_{\alpha,\ell}$ and Young's inequality, we conclude

$$\left| \xi^\alpha D_\xi^\alpha \widetilde{m}_{\kappa,\lambda}(\eta,\xi) \right| \leq c_0 \frac{|\kappa\eta| + |\xi|^2}{|N_{\kappa,\lambda}(\xi,\eta)|} \sum_{\ell=0}^{|\alpha|} \frac{|\xi|^{2|\alpha|}}{|N_{\kappa,\lambda}(\xi,\eta)|^\ell |\xi|^{2|\alpha|-2\ell}} \leq P_1(\lambda^2/\kappa)$$

by (A.89), where P_1 is a polynomial. To compute the derivative of $\widetilde{m}_{\kappa,\lambda}$ with respect to ξ_1, we differentiate each term of the sum in (A.88) separately and obtain

$$\partial_{\xi_1} \left[\frac{p_{\alpha,\ell}(\xi)}{N_{\kappa,\lambda}(\xi,\eta)^{\ell+1} |\xi|^{2|\alpha|-2\ell-2+2\theta}} \right] = \frac{\partial_{\xi_1} p_{\alpha,\ell}(\xi) |\xi|^2}{N_{\kappa,\lambda}(\xi,\eta)^{\ell+1} |\xi|^{2|\alpha|+2-2\ell-2+2\theta}}$$
$$- \frac{p_{\alpha,\ell}(\xi)(\ell+1)(2\xi_1 - i\lambda)}{N_{\kappa,\lambda}(\xi,\eta)^{\ell+2} |\xi|^{2|\alpha|-2\ell-2+2\theta}} - \frac{p_{\alpha,\ell}(\xi)(2|\alpha| - 2\ell - 2 + 2\theta)\xi_1}{N_{\kappa,\lambda}(\xi,\eta)^{\ell+1} |\xi|^{2|\alpha|+2-2\ell-2+2\theta}}.$$

Proceeding as above, we thus conclude the estimate

$$\left| \xi_1 \xi^\alpha \partial_{\xi_1} D_\xi^\alpha \widetilde{m}_{\kappa,\lambda}(\eta,\xi) \right|$$
$$\leq c_1 \frac{(|\kappa\eta| + |\xi|^2)(|\xi|^2 + |\lambda\xi_1|)}{|N_{\kappa,\lambda}(\xi,\eta)|^2} \sum_{\ell=0}^{|\alpha|} \frac{|\xi|^{2|\alpha|}}{|N_{\kappa,\lambda}(\xi,\eta)|^\ell |\xi|^{2|\alpha|-2\ell}} \leq P_2(\lambda^2/\kappa)$$

for a polynomial P_2. Next we compute the derivative of $\widetilde{m}_{\kappa,\lambda}$ with respect to η. Identity (A.88) yields

$$\partial_\eta D_\xi^\alpha \widetilde{m}_{\kappa,\lambda}(\eta,\xi) = \left((1 - \chi(\eta))|\eta|^{-2}\eta + \chi'(\eta) \right) \sum_{\ell=0}^{|\alpha|} \frac{|\kappa\eta|^\theta |\xi|^{2-2\theta} p_{\alpha,\ell}(\xi)}{N_{\kappa,\lambda}(\xi,\eta)^{\ell+1} |\xi|^{2|\alpha|-2\ell}}$$
$$- (1 - \chi(\eta)) \sum_{\ell=0}^{|\alpha|} \frac{|\kappa\eta|^\theta (\ell+1) i\kappa |\xi|^{2-2\theta} p_{\alpha,\ell}(\xi)}{N_{\kappa,\lambda}(\xi,\eta)^{\ell+2} |\xi|^{2|\alpha|-2\ell}}.$$

Employing $\chi'(\eta)|\eta| \leq 1$, we thus deduce

$$|\eta\xi^\alpha\partial_\eta D_\xi^\alpha \widetilde{m}_{\kappa,\lambda}(\eta,\xi)| \leq c_2\left(1 + \frac{|\kappa\eta|}{|N_{\kappa,\lambda}(\xi,\eta)|}\right)\sum_{\ell=0}^{|\alpha|}\frac{|\kappa\eta|^\theta|\xi|^{2-2\theta}|p_{\alpha,k}(\xi)||\xi|^{|\alpha|}}{|N_{\kappa,\lambda}(\xi,\eta)|^{\ell+1}|\xi|^{2|\alpha|-2\ell}}.$$

With the same argument as above, we see that the remaining terms are also bounded by $P_3(\lambda^2/\kappa)$ for a polynomial P_3. In the very same way, we derive a similar bound for $\eta\xi_1\xi^\alpha\partial_\eta\partial_{\xi_1}D_\xi^\alpha\widetilde{m}_{\kappa,\lambda}$. In total, we thus conclude the existence of a polynomial P_4 such that

$$|\eta^\beta\xi^\alpha\partial_\eta^\beta D_\xi^\alpha\widetilde{m}_{\kappa,\lambda}(\eta,\xi)| \leq P_4(\lambda^2/\kappa)$$

for all $\alpha \in \{0,1\}^n$, $\beta \in \{0,1\}$ and $(\eta,\xi) \in \mathbb{R} \times \mathbb{R}^n$ with $|\xi| \neq 0$. Now the Marcinkiewicz Multiplier Theorem (Theorem A.3.3), implies that $\widetilde{m}_{\kappa,\lambda}$ is an $L^q(\mathbb{R} \times \mathbb{R}^n)$ multiplier for all $q \in (1,\infty)$ and satisfies (A.90). $\qquad\square$

Bibliography

[1] M. Abramowitz and I. A. Stegun, editors. *Handbook of mathematical functions with formulas, graphs, and mathematical tables. 10th printing, with corrections.* New York: John Wiley & Sons, 1972. 170

[2] C. Amrouche, M. Meslameni, and Š. Nečasová. The stationary Oseen equations in an exterior domain: An approach in weighted Sobolev spaces. *J. Differential Equations*, 256(6):1955–1986, 2014. 10, 44

[3] C. Amrouche and U. Razafison. On the Oseen problem in three-dimensional exterior domains. *Anal. Appl. (Singap.)*, 4(2):133–162, 2006. 48

[4] W. Arendt and S. Bu. The operator-valued Marcinkiewicz multiplier theorem and maximal regularity. *Math. Z.*, 240(2):311–343, 2002. 6

[5] K. Babenko. On stationary solutions of the problem of flow past a body of a viscous incompressible fluid. *Math. USSR, Sb.*, 20:1–25, 1973. 15

[6] K. I. Babenko and M. M. Vasil'ev. On the asymptotic behavior of a steady flow of viscous fluid at some distance from an immersed body. *Prikl. Mat. Meh.*, 37:690–705, 1973. In Russian. English translation: *J. Appl. Math. Mech.*, 37:651–665, 1973. 15, 148

[7] H. Brézis and L. Nirenberg. Forced vibrations for a nonlinear wave equation. *Commun. Pure Appl. Math.*, 31(1):1–30, 1978. 5

[8] F. E. Browder. Existence of periodic solutions for nonlinear equations of evolution. *Proc. Natl. Acad. Sci. U.S.A.*, 53(5):1100–1103, 1965. 4

[9] F. Bruhat. Distributions sur un groupe localement compact et applications à l'étude des représentations des groupes p-adiques. *Bull. Soc. Math. Fr.*, 89:43–75, 1961. 23

[10] V. I. Burenkov. Extension of functions with preservation of the Sobolev seminorm. *Trudy Mat. Inst. Steklov.*, 172:71–85, 1985. (English transl.: *Proc. Steklov Inst. Math.*, 3:81–95, 1987). 35

[11] A. Celik and M. Kyed. Nonlinear Wave Equation with Damping: Periodic Forcing and Non-Resonant Solutions to the Kuznetsov Equation. *ZAMM Z. Angew. Math. Mech.*, 98(3):412–430, 2018. 6

[12] A. Celik and M. Kyed. Nonlinear acoustics: Blackstock-Crighton equations with a periodic forcing term. *J. Math. Fluid Mech.*, 21(3), 2019. 6

[13] L. Cesari. Existence in the large of periodic solutions of hyperbolic partial differential equations. *Arch. Ration. Mech. Anal.*, 20(3):170–190, 1965. 5

[14] D. C. Clark. The vorticity at infinity for solutions of the stationary Navier-Stokes equations in exterior domains. *Indiana Univ. Math. J.*, 20:633–654, 1970/71. 15, 148

[15] F. Crispo and P. Maremonti. An interpolation inequality in exterior domains. *Rend. Sem. Mat. Univ. Padova*, 112:11–39, 2004. 31

[16] K. de Leeuw. On L_p multipliers. *Ann. of Math. (2)*, 81:364–379, 1965. 25

[17] P. Deuring and G. P. Galdi. Exponential decay of the vorticity in the steady-state flow of a viscous liquid past a rotating body. *Arch. Ration. Mech. Anal.*, 221(1):183–213, 2016. 15, 136

[18] R. Edwards and G. Gaudry. *Littlewood-Paley and multiplier theory*. Berlin-Heidelberg-New York: Springer-Verlag, 1977. 25, 26

[19] T. Eiter and G. P. Galdi. New results for the Oseen problem with applications to the Navier-Stokes equations in exterior domains. arXiv:1904.01527, 2019. 11, 43

[20] T. Eiter and M. Kyed. Time-periodic linearized Navier-Stokes Equations: An approach based on Fourier multipliers. In *Particles in Flows*, Adv. Math. Fluid Mech., pages 77–137. Birkhäuser/Springer, Cham, 2017. 23, 26

[21] T. Eiter and M. Kyed. Estimates of time-periodic fundamental solutions to the linearized Navier-Stokes equations. *J. Math. Fluid Mech.*, 20(2):517–529, 2018. 14, 105, 118, 121

[22] T. Eiter and M. Kyed. Viscous flow around a rigid body performing a time-periodic motion. arXiv:1912.04938, 2019. 13

[23] T. Eiter, M. Kyed, and Y. Shibata. $\mathcal{R}$-solvers and their application to periodic L^p estimates. (In preparation). 5

[24] T. Eiter, M. Kyed, and Y. Shibata. Falling drop in an unbounded liquid reservoir: Steady-state solutions. arXiv:1912.04925, 2019. 12

[25] T. Eiter, M. Kyed, and Y. Shibata. On periodic solutions for one-phase and two-phase problems of the Navier–Stokes equations. arXiv:1909.13558, 2019. 12

[26] R. Farwig. Das stationäre Außenraumproblem der Navier-Stokes-Gleichungen bei nichtverschwindender Anströmgeschwindigkeit in anisotrop gewichteten Sobolevräumen. SFB 256 preprint no. 110 (Habilitationsschrift). University of Bonn (1990). 116, 117

[27] R. Farwig. The stationary exterior 3D-problem of Oseen and Navier-Stokes equations in anisotropically weighted spaces. *Math. Z.*, 211(3):409–448, 1992. 117

[28] R. Farwig, G. P. Galdi, and M. Kyed. Asymptotic structure of a Leray solution to the Navier-Stokes flow around a rotating body. *Pac. J. Math.*, 253(2):367–382, 2011. 15

[29] R. Farwig and T. Hishida. Leading term at infinity of steady Navier-Stokes flow around a rotating obstacle. *Math. Nachr.*, 284(16):2065–2077, 2011. 15

[30] R. Farwig, T. Hishida, and D. Müller. L^q-theory of a singular winding integral operator arising from fluid dynamics. *Pac. J. Math.*, 215(2):297–312, 2004. 104

[31] R. Farwig and J. Neustupa. On the spectrum of an Oseen-type operator arising from flow past a rotating body. *Integral Equations Operator Theory*, 62:169–189, 2008. 75

[32] R. Farwig and J. Neustupa. Spectral properties in L^q of an Oseen operator modelling fluid flow past a rotating body. *Tohoku Math. J. (2)*, 62(2):287–309, 2010. 75

[33] R. Farwig and H. Sohr. Generalized resolvent estimates for the Stokes system in bounded and unbounded domains. *J. Math. Soc. Japan*, 46(4):607–643, 1994. 40, 86

[34] F. A. Ficken and B. A. Fleishman. Initial value problems and time-periodic solutions for a nonlinear wave equation. *Commun. Pure Appl. Math.*, 10(3):331–356, 1957. 3, 5

[35] R. Finn. An energy theorem for viscous fluid motions. *Arch. Ration. Mech. Anal.*, 6:371–381, 1960. 138

[36] R. Finn. On the exterior stationary problem for the Navier-Stokes equations, and associated perturbation problems. *Arch. Ration. Mech. Anal.*, 19:363–406, 1965. 15

[37] E. Gagliardo. Ulteriori proprietà di alcune classi di funzioni in più variabili. *Ricerche Mat.*, 8:24–51, 1959. 30

[38] G. Galdi and H. Sohr. Existence and uniqueness of time-periodic physically reasonable Navier-Stokes flow past a body. *Arch. Ration. Mech. Anal.*, 172(3):363–406, 2004. 8

[39] G. P. Galdi. On the asymptotic structure of D-solutions to steady Navier-Stokes equations in exterior domains. In *Mathematical problems relating to the Navier-Stokes equation*, volume 11 of *Ser. Adv. Math. Appl. Sci.*, pages 81–104. World Sci. Publ., River Edge, NJ, 1992. 15

[40] G. P. Galdi. On the Oseen boundary value problem in exterior domains. In *The Navier-Stokes equations II—theory and numerical methods (Oberwolfach, 1991)*, volume 1530 of *Lecture Notes in Math.*, pages 111–131. Springer, Berlin, 1992. 10, 44, 46, 48

[41] G. P. Galdi. On the motion of a rigid body in a viscous liquid: a mathematical analysis with applications. In *Handbook of mathematical fluid dynamics, Vol. I*, pages 653–791. North-Holland, Amsterdam, 2002. 7

[42] G. P. Galdi. *An introduction to the mathematical theory of the Navier-Stokes equations. Steady-state problems. 2nd ed.* New York: Springer, 2011. 10, 17, 28, 36, 37, 44, 46, 47, 48, 50, 55, 77, 79, 95, 108, 109, 117, 139, 166

[43] G. P. Galdi. Existence and uniqueness of time-periodic solutions to the Navier-Stokes equations in the whole plane. *Discrete Contin. Dyn. Syst. Ser. S*, 6(5):1237–1257, 2013. 6, 8

[44] G. P. Galdi. On time-periodic flow of a viscous liquid past a moving cylinder. *Arch. Ration. Mech. Anal.*, 210(2):451–498, 2013. 6, 8

[45] G. P. Galdi. A Time-Periodic Bifurcation Theorem and its Application to Navier-Stokes Flow Past an Obstacle. In *Mathematical Analysis of Viscous Incompressible Fluid*, number 1971 in RIMS Kôkyûroku, pages 1–27. Res. Inst. Math. Sci. (RIMS), Kyoto, 2014. 11

[46] G. P. Galdi. On bifurcating time-periodic flow of a Navier-Stokes liquid past a cylinder. *Arch. Ration. Mech. Anal.*, 222(1):285–315, 2016. 11

[47] G. P. Galdi. Viscous flow past a body translating by time-periodic motion with zero average. arXiv:1903.03840, 2019. 8

[48] G. P. Galdi and M. Kyed. A simple proof of L^q-estimates for the steady-state Oseen and Stokes equations in a rotating frame. Part I: Strong solutions. *Proc. Amer. Math. Soc.*, 141(2):573–583, 2013. 13

[49] G. P. Galdi and M. Kyed. A simple proof of L^q-estimates for the steady-state Oseen and Stokes equations in a rotating frame. Part II: Weak solutions. *Proc. Amer. Math. Soc.*, 141(4):1313–1322, 2013. 13

[50] G. P. Galdi and M. Kyed. Time-periodic flow of a viscous liquid past a body. In *Partial differential equations in fluid mechanics*, volume 452 of *London Math. Soc. Lecture Note Ser.*, pages 20–49. Cambridge Univ. Press, Cambridge, 2018. 8, 11, 32, 33, 55, 57, 127

[51] G. P. Galdi and M. Kyed. Time-periodic solutions to the Navier-Stokes equations in the three-dimensional whole-space with a non-zero drift term: Asymptotic profile at spatial infinity. In *Mathematical Analysis in Fluid Mechanics: Selected Recent Results*, volume 710

of *Contemp. Math.*, pages 121–144. Amer. Math. Soc., Providence, RI, 2018. 14, 15, 16, 136, 138, 139, 140

[52] G. P. Galdi and A. L. Silvestre. Strong solutions to the Navier-Stokes equations around a rotating obstacle. *Arch. Ration. Mech. Anal.*, 176(3):331–350, 2005. 88

[53] G. P. Galdi and A. L. Silvestre. Existence of time-periodic solutions to the Navier-Stokes equations around a moving body. *Pac. J. Math.*, 223(2):251–267, 2006. 8

[54] G. P. Galdi and A. L. Silvestre. On the motion of a rigid body in a Navier-Stokes liquid under the action of a time-periodic force. *Indiana Univ. Math. J.*, 58(6):2805–2842, 2009. 8

[55] M. Geissert, M. Hieber, and T. H. Nguyen. A general approach to time periodic incompressible viscous fluid flow problems. *Arch. Ration. Mech. Anal.*, 220(3):1095–1118, 2016. 9

[56] L. Grafakos. *Classical Fourier analysis. 2nd ed.* New York, NY: Springer, 2008. 177, 178

[57] L. Grafakos. *Modern Fourier analysis. 2nd ed.* New York, NY: Springer, 2009. 34, 177

[58] W. S. Hall. Periodic solutions of a class of weakly nonlinear evolution equations. *Arch. Ration. Mech. Anal.*, 39:294–322, 1970. 5

[59] J. G. Heywood. The Navier-Stokes equations: On the existence, regularity and decay of solutions. *Indiana Univ. Math. J.*, 29:639–681, 1980. 41

[60] E. Hopf. Über die Anfangswertaufgabe für die hydrodynamischen Grundgleichungen. *Math. Nachr.*, 4:213–231, 1951. 4

[61] S. Ibrahim, P. G. Lemarié-Rieusset, and N. Masmoudi. Time-periodic forcing and asymptotic stability for the Navier-Stokes-Maxwell equations. *Commun. Pure Appl. Math.*, 71(1):51–89, 2018. 6

[62] K. Kang, H. Miura, and T.-P. Tsai. Asymptotics of small exterior Navier-Stokes flows with non-decaying boundary data. Preprint. 15

[63] S. Kaniel and M. Shinbrot. A reproductive property of the Navier-Stokes equations. *Arch. Rational Mech. Anal.*, 24:363–369, 1967. 5

[64] D. Kim and H. Kim. L^q-estimates for the stationary Oseen equations on exterior domains. *J. Differential Equations*, 257(10):3669–3699, 2014. 48, 49

[65] Y. S. Kolesov. Certain tests for the existence of stable periodic solutions of quasi-linear parabolic equations. *Sov. Math. Dokl.*, 5:1118–1120, 1964. 4

[66] Y. S. Kolesov. A test for the existence of periodic solutions to parabolic equations. *Sov. Math. Dokl.*, 7:1318–1320, 1966. 4

[67] Y. S. Kolesov. Periodic solutions of second order quasilinear parabolic equations. *Trudy Moskov. Mat. Obšč.*, 21:103–134, 1970. 4

[68] A. Korolev and V. Šverák. On the large-distance asymptotics of steady state solutions of the Navier-Stokes equations in 3D exterior domains. *Ann. Inst. Henri Poincaré, Anal. Non Linéaire*, 28(2):303–313, 2011. 15

[69] H. Kozono and M. Nakao. Periodic solutions of the Navier-Stokes equations in unbounded domains. *Tohoku Math. J. (2)*, 48(1):33–50, 1996. 5, 7

[70] M. Krasnosel'skiĭ. The operator of translation along the trajectories of differential equations. *Translations of Mathematical Monographs*, 1968. 4

[71] S. Kračmar, A. Novotný, and M. Pokorný. Estimates of Oseen kernels in weighted L^p spaces. *J. Math. Soc. Japan*, 53(1):59–111, 2001. 117, 176

[72] M. Kyed. Time-Periodic Solutions to the Navier-Stokes Equations. *Habilitationsschrift, Technische Universität Darmstadt*, 2012. 6, 138

[73] M. Kyed. The existence and regularity of time-periodic solutions to the three-dimensional Navier-Stokes equations in the whole space. *Nonlinearity*, 27(12):2909–2935, 2014. 6, 8, 10

[74] M. Kyed. Maximal regularity of the time-periodic linearized Navier-Stokes system. *J. Math. Fluid Mech.*, 16(3):523–538, 2014. 6, 8, 126

[75] M. Kyed. On the asymptotic structure of a Navier-Stokes flow past a rotating body. *J. Math. Soc. Japan*, 66(1):1–16, 2014. 15

[76] M. Kyed. A fundamental solution to the time-periodic Stokes equations. *J. Math. Anal. Appl.*, 437(1):708–719, 2016. 14, 105, 118

[77] R. Larsen. *An introduction to the theory of multipliers.* Springer-Verlag, New York-Heidelberg, 1971. Die Grundlehren der mathematischen Wissenschaften, Band 175. 25

[78] J. Leray. Étude de diverses équations intégrales non linéaires et de quelques problèmes que pose l'hydrodynamique. *J. Math. Pures Appl.*, 12:1–82, 1933. 3

[79] J. Leray. Sur le mouvement d'un liquide visqueux emplissant l'espace. *Acta Math.*, 63(1):193–248, 1934. 3

[80] H. Lorentz. Eene algemeene stelling omtrent de beweging eener vloeistof met wrijving en eenige daaruit afgeleide gevolgen. *Verslagen der Afdeeling Natuurkunde van de Koninklijke Akademie van Wetenschappen*, 5:168–175, 1896. 14, 105

[81] P. Maremonti. Existence and stability of time-periodic solutions to the Navier-Stokes equations in the whole space. *Nonlinearity*, 4(2):503–529, 1991. 7

[82] P. Maremonti. Some theorems of existence for solutions of the Navier-Stokes equations with slip boundary conditions in half-space. *Ric. Mat.*, 40(1):81–135, 1991. 7

[83] P. Maremonti and M. Padula. Existence, uniqueness and attainability of periodic solutions of the Navier-Stokes equations in exterior domains. *Zap. Nauchn. Sem. S.-Peterburg. Otdel. Mat. Inst. Steklov. (POMI)*, 233(Kraev. Zadachi Mat. Fiz. i Smezh. Vopr. Teor. Funkts. 27):142–182, 257, 1996. 7

[84] L. Nirenberg. On elliptic partial differential equations. *Ann. Scuola Norm. Sup. Pisa Cl. Sci. (3)*, 13:115–162, 1959. 30

[85] C. W. Oseen. *Hydrodynamik.* Akademische Verlagsgesellschaft M.B.H., Leipzig, 1927. 14, 105

[86] H. Poincaré. *Sur le problème des trois corps et les équations de la dynamique*. Acta mathematica. F. & G. Beijer, 1890. 4

[87] H. Poincaré. *The three-body problem and the equations of dynamics. Poincaré's foundational work on dynamical systems theory. Translated from the French by Bruce D. Popp.*, volume 443. Cham: Springer, Originally published by Institut Mittag-Leffler, Sweden, 1890 edition, 2017. 4

[88] G. Prodi. Soluzioni periodiche di equazioni alle derivate parziali di tipo parabolico e non lineari. *Riv. Mat. Univ. Parma*, 3:265–290, 1952. 3, 5

[89] G. Prodi. Soluzioni periodiche di equazioni a derivate parziali di tipo iperbolico non lineari. *Ann. Mat. Pura Appl. (4)*, 42:25–49, 1956. 3, 5

[90] G. Prodi. Qualche risultato riguardo alle equazioni di Navier-Stokes nel caso bidimensionale. *Rend. Sem. Mat. Univ. Padova*, 30:1–15, 1960. 4

[91] G. Prouse. Soluzioni periodiche dell'equazione di Navier-Stokes. *Atti Accad. Naz. Lincei Rend. Cl. Sci. Fis. Mat. Natur. (8)*, 35:443–447, 1963. 4

[92] G. Prouse. Soluzioni quasi-periodiche dell'equazione differenziale di Navier-Stokes in due dimensioni. *Rend. Sem. Mat. Univ. Padova*, 33:186–212, 1963. 4

[93] P. H. Rabinowitz. Periodic solutions of nonlinear hyperbolic partial differential equations. *Commun. Pure Appl. Math.*, 20:145–205, 1967. 5

[94] P. H. Rabinowitz. Periodic solutions of nonlinear hyperbolic partial differential equations. II. *Commun. Pure Appl. Math.*, 22:15–39, 1969. 5

[95] W. Rudin. *Fourier analysis on groups*. Wiley Classics Library. John Wiley & Sons, Inc., New York, 1990. Reprint of the 1962 original, A Wiley-Interscience Publication. 17

[96] J. Serrin. A note on the existence of periodic solutions of the Navier-Stokes equations. *Arch. Ration. Mech. Anal.*, 3:120–122, 1959. 3, 5

[97] H. Sohr. *The Navier-Stokes equations. An elementary functional analytic approach.* Basel: Birkhäuser, 2001. 40

[98] I. Stakgold. *Boundary value problems of mathematical physics. Vol. I and II. Reprint of the 1967/68 originals.* Philadelphia, PA: Society for Industrial and Applied Mathematics (SIAM), 2000. 109

[99] A. Takeshita. On the reproductive property of the 2-dimensional Navier-Stokes equations. *J. Fac. Sci. Univ. Tokyo Sect. I*, 16:297–311 (1970), 1969. 5

[100] A. E. Taylor and D. C. Lay. *Introduction to functional analysis.* John Wiley & Sons, New York-Chichester-Brisbane, second edition, 1980. 88

[101] L. Weis. Operator-valued Fourier multiplier theorems and maximal L_p-regularity. *Math. Ann.*, 319(4):735–758, 2001. 6

[102] M. Yamazaki. The Navier-Stokes equations in the weak-L^n space with time-dependent external force. *Math. Ann.*, 317(4):635–675, 2000. 5, 7

[103] V. I. Yudovich. Periodic motions of a viscous incompressible fluid. *Sov. Math. Dokl.*, 1:168–172, 1960. 4

Index

Curriculum Vitae

08/22/91   Born in Gelnhausen, Germany

08/02 – 06/11   **Secondary school,** *Ulrich-von-Hutten-Gymnasium,* Schlüchtern, Germany, Abitur (1.1 / very good)

10/11 – 10/14   **Bachelor's studies,** *Technische Universität Darmstadt,* Darmstadt, Germany, Bachelor of Science Mathematics (1.24 / very good)
Bachelor's thesis: *Der Satz von Mihlin und $\mathcal{H}^\infty$-Kalkül für elliptische Operatoren*

10/14 – 9/16   **Master's studies,** *Technische Universität Darmstadt,* Darmstadt, Germany, Master of Science Mathematics (1.10 / very good)
Master's thesis: *Fourier Multipliers on Locally Compact Abelian Groups and their Applications to Partial Differential Equations*

10/16 – 02/20   **Doctoral studies,** *Technische Universität Darmstadt,* Darmstadt, Germany, research assistant in the working group *Partielle Differentialgleichungen*

12/12/19   **Submission of the doctoral thesis (Dissertation),** *Existence and Spatial Decay of Periodic Navier–Stokes Flows in Exterior Domains* at *Technische Universität Darmstadt,* Darmstadt, Germany

02/27/20   **Defense of the doctoral thesis,** overall assessment *summa cum laude*